MW01309187

Yoga and Ecology

Dharma for the Earth

Contemporary Issues in Constructive Dharma Vol. 6
Proceedings of Sessions of the Fourth DANAM Conference

DEEPAK HERITAGE BOOKS
Published Volumes and Volumes in Preparation

Indic Heritage Series

Holy War: Violence and the Bhagavad Gita
Steven J. Rosen (Ed.)

Journal of Vaishnava Studies
Senior Editors: Steven J. Rosen and Graham M. Schweig

10th Anniversary Journal of Vaishnava Studies CD
Graham M. Schweig (Ed.)

Contemporary Issues in Constructive Dharma (CICD) Series
*Contemporary Issues in Constructive Dharma Vol. I**
(*on Science and Representation in Hindu Traditions)
Rita D. Sherma and Adarsh Deepak (Eds.)

Contemporary Issues in Constructive Dharma Vol. II
Hermeneutics and Epistemology
Rita D. Sherma and Adarsh Deepak (Eds.)

Contemporary Issues in Constructive Dharma Vol. 3
Asceticism, Identity, and Pedagogy in Dharma Traditions
Graham M. Schweig, Jeffery D. Long, Ramdas Lamb and Adarsh Deepak (Eds.)

Contemporary Issues in Constructive Dharma Vol. 4
Death and Afterlife in Dharma Traditions and Western Religions
Adarsh Deepak and Rita D. Sherma (Eds.)

Contemporary Issues in Constructive Dharma Vol. 5
Authority and Its Challenges in Hindu Texts, Translations, and Transnational Communities
Deepak Sarma (Ed.)

Contemporary Issues in Constructive Dharma Vol. 6
Yoga and Ecology: Dharma for the Earth
Christopher Key Chapple (Ed.)

Yoga and Ecology
Dharma for the Earth

Editor

Christopher Key Chapple
Doshi Professor of Indic and Comparative Theology
Loyola Marymount University

Contemporary Issues in Constructive Dharma Vol. 5
Proceedings of Two of the Sessions at the Fourth DANAM Conference
Held on site at the American Academy of Religion
Washington, DC, 17–19 November 2006

2009

Deepak Heritage Books
Hampton, Virginia, USA
www.deepakheritage.com

DEEPAK HERITAGE BOOKS

Published by

A. Deepak Publishing
A division of Science and Technology Corporation
10 Basil Sawyer Drive
Hampton, Virginia 23666
www.deepakheritage.com

Copyright © 2007 by A. Deepak Publishing
All rights reserved.
ISBN: 978-0-937194-53-9

Cover design:

The theme elements of the cover design and the DANAM Logo [composed of stylized 'infinite wave', and the acronym DANAM (pronounced dā, as in cart + nam, as in number) in both English and Sanskrit (inside of the oval in devanagari script)] are taken, with permission of Dharma Association of North America, from the DANAM web site design at www.danam-web.org. The 'infinite wave' symbolically represents Dharma, a term (with no English equivalent) derived from the Sanskrit (Sk.) root, dhri, "to sustain", that implies innate or natural order, law and code sustaining the changing world (jagat, in Sk.) of both non-living and living matter, undergoing change in a periodic manner, without beginning and end. Dharma is often traditionally called Sanātana Dharma, the Sk. term 'sanātana' (eternal and imperishable) being a qualifier, rather than an adjective.

Library of Congress Cataloging-in-Publication Data

DANAM Conference (4th : 2006 : Washington, D.C.)
Yoga and ecology : Dharma for the earth : proceedings of two of the sessions at the Fourth DANAM Conference, held on site at the American Academy of Religion, Washington, DC, 17–19 November 2006 / editor, Christopher Key Chapple.
p. cm. — (Contemporary issues in constructive Dharma ; v. 6)
Includes bibliographical references and index.
ISBN 978-0-937194-53-9 (alk. paper)
1. Yoga—Congresses. 2. Ecology—Congresses. I. Chapple, Christopher Key, 1954- II. Title.
B132.Y6D15 2006
181′.45—dc22 2009036766

Printed in the United States of America

CONTENTS

Page

Acknowledgements vii

Introduction 1
Christopher Key Chapple

Yoga and Nature: Vital Concentration in Atharva Veda 13
Suzanne Ironbiter

Pṛthivī Sūkta: Earth Verses 23
Christopher Key Chapple

The Integration of Spirit (Puruṣa) and Matter (Prakṛti) in the Yoga Sūtra 33
Ian Whicher

Super Saṃskāras: Soteriological Subliminal Impressions in Patañjali's Yoga Sūtra 61
Beverley Foulks

Dharmamegha Samādhi and the Two Sides of Kaivalya: Toward a Yogic Theory of Culture 83
Alfred Collins

Toward a Yoga-Inspired Environmental Ethics 97
Christopher Key Chapple

The Disharmony of Interdependence: Sāṃkhya-Yoga and Ecology 105
Knut A. Jacobsen

Towards a Theory of Tantra-Ecology 123
Jeffrey S. Lidke

Green Yoga: Contemporary Activism and Ancient Practices: A Model for Eight Paths of Green Yoga 145
Laura Cornell

Blibliography 171

Contributors 189

Index 193

ACKNOWLEDGEMENTS

Several persons helped inspire the development of this book, including Adarsh Deepak, founder and President of the Dharma Association of North America (DANAM), sponsor of the 2006 gathering in Washington, D.C., where these papers were first presented. Special thanks are due members of the DANAM Board of Trustees (Dr. Rajinder Gandhi, MD, DANAM Chairman) for their generous financial support. The editor wishes to acknowledge the enthusiastic support and cooperation of the members of the program committee, session conveners (Jeffrey C. Ruff, Marshall University, and Christopher Chapple, Loyola Marymount University), and presiders, (Ramdas Lamb, University of Hawaii at Manoa, and Daniel Michon, Loyola Marymount University), presenters and participants for making this a stimulating conference session for everyone.

Co-Chair Rita DasGupta Sherma of Binghamton University and Stuart Sarbacker of Northwestern University offered key support. Mary Evelyn Tucker and John Grim of Yale University, founders of the Forum of Religion and Ecology, have encouraged the work of the Green Yoga Association since its inception. Likewise, Bron Taylor of the University of Florida and founder of the Society for the Study of Religion, Nature, and Culture, has been very supportive of this field of inquiry.

It is a pleasure to acknowledge the valuable assistance of Diana McQuestion and Delores Shackelford of Deepak Heritage Books, and especially of Nicole de Picciotto and Viresh Hughes in the preparation of this manuscript and production of the book.

Christopher Key Chapple
Editor

INTRODUCTION

Christopher Key Chapple

This book seeks to provide four points of connection that can help elucidate the ways in which the ancient and continuing tradition of Yoga might prove a relevant and helpful dialogue partner as the religious philosophies of the world grapple with the looming threat of climate change, species decimation, and an overall diminishment of the prospects for a healthy planet. This first point of connection reaches back into the pre-articulated ground that helped nourish the growth and development of Yoga, namely the Vedic tradition. The second area to be discussed will explore whether classical Yoga, generally associated with the abnegation of all things worldly, holds a foundational interest in the material world. Tantra, the third area of exploration, suggests that the human body and the engagement of the senses can be employed for the purposes of enjoyment in the service of the spiritual values set forth by Yoga. Finally, the book concludes with contemporary approaches of Yoga by theorists and activists who have directly employed Yoga methodologies to advance the cause of heightening ecological sensitivity.

Some scholars have disputed the applicability of Yoga as a valid mode of discourse for dealing with environmental issues, most notably Lance Nelson and Rita Sherma. Nelson writes that nature is irrelevant to spiritual life, "ultimately unimportant (*tuccha*)".[1] He also suggests that the *Bhagavad Gītā*, one of the authoritative texts on the practice of Yoga, "teaches that our true self is outside the world of nature"[2] and hence not a good resource for developing an environmental ethic or even cultivating ecological sensitivity. Sherma contends that due to its association with women, nature has been "enmeshed in a net of devaluation," resulting in "pervasive pollution" throughout India.[3] She claims that both widely known terms for the feminine creative force, Prakṛti and Māyā, are obstacles to be overcome, not occasions for celebration.[4] According to Sherma, both Sāṃkhya and Vedānta must be rejected as ecologically friendly because of their emphasis on transcendence and their devaluation of the created world unless re-interpreted through the prism of Tantra, which in her assessment re-valorizes the feminine and hence nature herself. Similarly, J. Baird Callicott states that "Hindu religious practice seeks to transcend this world, not improve it"[5] and finds Hinduism "essentially hostile to environmental concerns".[6] However, by looking back to the early origins of Yoga, by examining its key ideas as articulated in Patañjali, and taking into account both Tantra Yoga and the worldwide phenomenon of Modern Yoga, this book aims to suggest that Yoga and by extension the Hindu tradition, may be seen as holding resources for the conceptualization and actualization of environmental awareness and ethics.

A Brief History of Yoga

Although some may dispute the validity of non-textual artifacts to establish the origins of Yoga, possible early evidence of Yoga practice can be found in Indus Valley seals unearthed in Mohenjodaro and other cities that date from 3500 B.C.E. These images seems to depict men and women in ritual or meditative poses, often surrounded with an array of plant life and animals, domestic and wild. These same images, generally referred to as part of the Paśupati or Lord of Animals motif, appear in medieval temples, indicating a continuity of culture for over five thousand years. Textual references to Yoga appear in the middle *Upaniṣads* such as the *Maitri* and the *Śvetāśvatara* and in the massive epic text, the *Mahābhārata*, including the segment known as the *Bhagavad Gītā*, sometimes referred to as the "Hindu Bible." These materials date from about 600 B.C.E. The Buddha and the Jina (ca. 500 B.C.E.) both taught yogic styles of meditation as well as ethical systems that share a concern for restraint from violence. By around 200 C.E., Patañjali summarized Yoga practices in a classic text known as the *Yoga Sūtra*. Sanskrit texts such as the *Yogavāsiṣṭha* (ca. 1000 C.E.) and the *Hatha Yoga Pradīpikā* (ca. 1500 C.E.) describe various forms of Vedantic and esoteric Yoga. Haribhadra's *Yogadṛṣṭisamuccya* (ca. 750 C.E.) and Hemacandra's *Yogaśāstra* (ca. 1250 C.E.) discuss the adaptation of the tradition to the Jaina faith, while the later texts of Kabir and Guru Nanak allude to Yoga meditation from universalist and Sikh perspectives. Yoga continues to be practiced throughout India and has become increasingly popular worldwide.

The Vedic Context

Yoga arose in India, on a subcontinent delineated by vast oceans to the west, south, and east, and the lofty Himalayas to the north. The Vedas, the earliest texts of India, celebrate the geography of India with great enthusiasm. These texts were orally transmitted for several hundred years before the advent of writing, and scholars have estimated that they were composed between 1500 to 900 B.C.E. Many Vedic hymns personify and extol the rhythm of seasonal changes, with the dragon Vṛtra symbolizing the long period of drought and heat, and the god Indra wielding his thunderbolt to release the monsoon rains. These early hymns chart a delicate and intimate relationship between humans and the cosmos, and represent some of earliest articulations of how the microphase reflects the macrophase. Suzanne Ironbiter examines this connection between the lower domains and the great above as an early modeling of environmental and ecological awareness. Chapple's first essay explores verses from the *Pṛthivī Sūkta* or Earth Verses of the *Atharva Veda* that offer praise and reverence to the earth. Though Yoga does not find direct mention in the Vedic materials, the Vedic worldview helps define the context for Yoga as it develops.

Classical Sāṃkhya and Yoga

In distinction to schools of Indian thought that refer to the world as illusory, Yoga asserts the reality of the world. It builds upon the Sāṃkhya school of philosophy, first espoused by a mythical sage named Kapila who perhaps lived in northeastern India around 900 B.C.E. Kapila's teachings were later systematized by a philosopher known as Īśvarakrishna, who composed the *Sāṃkhya Kārikā* in the early centuries of the common era. In this seminal text, the author exerts great care to articulate the existence and importance of the natural world. He posits that the world is known to us through its effects, and the effects stem from a common cause, *prakṛti*, a term that many scholars choose to translate as "nature." This realism, known in Sanskrit as *satkaryavāda* or "things originate from a pre-existent, enduring nature," claims that the soul, often referred to elliptically as an unknowable consciousness, and the things of the world rely upon two eternal principles. Nature (*prakṛti*) provides experience and liberation for her silent observer, the spiritual consciousness or *puruṣa*. According to the *Sāṃkhya Kārikā*, all things exist for the purpose of serving and liberating this consciousness. Through understanding the creative force known as nature, one advances toward a state of freedom. Through understanding the structure and purpose of things, one is able to cultivate a state of nonattachment that, from the perspective of this philosophy, entails a state of appreciation and respect for nature, not disdain and abnegation.

Ian Whicher suggests that the culmination of Yoga in a state of blessed solitude (*kaivalyam*) redeems the world; in the past, others have seen this solitude as a rebuff of the world. Beverly Foulks, in partial agreement with Whicher, suggests that a more valid case for worldly affirmation can be made by examining the remnants of karma (*saṃskāras*) that allow one to engage the world through a place of ritual purity rather than attachment. Alfred Collins sees the description of *Dharma Megha Samādhi* as affirming the paradox that the world, even in its evanescence, can manifest an enlightened view. All three arguments suggest that classical Yoga can be used to affirm the overall task of environmental ethics and need not be seen as an escape from or denial of conventional realities.

In the middle Upaniṣadic period of Yoga, we find speculative discourses and dialogues about the nature and function of the human body and mind. By reflecting on the functions of the body, particularly the breath, and by seeking to still the mind, the Upaniṣads state that one can establish a connection with one's inner self or Atman, often translated as soul. Passages from the early Upaniṣads such as the *Chāndogya* and *Bṛhadāraṇyaka Upaniṣads* emphasize the primacy of breath and the relationship between the microphase and the macrophase aspects of reality. By getting to know oneself through focusing on the power of the breath, one feels an intimacy with the larger aspects of the earth and heavens, perhaps most aptly conveyed in the first section of the *Bṛhadāraṇyaka Upaniṣad*, which first correlates the various functions and regions of the universe with the cosmic horse, and then makes a similar series of correspondences with the human body. By understanding one's desires and impulses, as well as the structures and functions of one's body and mind, one gains an understanding of the cosmos.

The later Upaniṣads and the *Bhagavad Gītā* speak directly of Yoga as the technique to be utilized in order to feel that intimate connection with the flow of life and one's place within reality. The *Śvetāśvatara* and *Maitri Upaniṣads* state that by drawing the senses inward and controlling the breath, one can reach a state of equipoise. The *Bhagavad Gītā* comes to describe the Yogi as one who comprehends the relationship between the "field" or nature (*prakṛti*) and the "knower of the field" or spirit (*puruṣa*). Within the body of Krishna, the entire world, in its splendor and terror, can be seen, appreciated, and embraced. The metaphor of the human body becomes extended in the *Gītā* to include all aspects of the universe.

Tantra

The Upaniṣadic philosophy of correlations between microphase and macrophase found popular expression in the systems of Tantra that began to find wide notice in India by the seventh century of the common era. In Tantra, the divine feminine rises to prominence as the matrix for all experience. Yoga practices as employed by practitioners of Tantra enhance one's control over the material world. Central to the process of cosmogenesis espoused in Tantra is the notion that the human body holds the key to spiritual realization. Lidke's chapter provides an overview of this tradition, suggesting that the insights of Tantra might be particularly instructive for developing a workable ecological lifestyle.

Modern Yoga

Most contemporary expressions of Yoga hearken back to the *Bhagavad Gītā* and Patañjali's *Yoga Sūtra* as authoritative sources for interpreting Yoga. Knut Jacobsen shows how Arne Naess, the founder of the Deep Ecology movement, found inspiration in the life and work of Mahatma Gandhi, and particularly his interpretation of the *Bhagavad Gītā.* Laura Cornell, founder of the Green Yoga Association, shows connections between the Yogas of the *Bhagavad Gītā*, the practices of Yoga as articulated by Patañjali, and contemporary forms of Yoga that are being developed in light of environmental ethics and activism.

This collection of essays suggests that the rejection of the world normally associated with India's traditions of renunciation can be reconfigured in such a way that its core message can be used for transformation rather than transcendence. Stephanie Kaza, in her book *Hooked: Desire and Greed* has provided a critique of consumerism as the root cause of environment suffering. She notes that: "The [Buddhist] precepts represent practices of restraint, calling for personal responsibility for reducing environmental and human suffering...Because social structures (governments, schools, churches, and so on) contribute to consumer-related harming, ethical guidelines for social structures would also be useful...[so] consumers could reclaim moral integrity that has been eroded by consumerist agendas" (pp. 146–148). Similarly, just as Buddhism has proven instructive in regard to a critique of materialism, so also the basic truths of Yoga can be applied to envision an emerging environmental sensitivity.

Yoga in the Context of Religion and Ecology

Several scholars, most notably Lynn White, Jr., have suggested that religion has contributed toward the problem of environmental decline in part due to its narrative that places the human order in a dominium relationship over nature. His essay has sparked a long debate on this issue with many theologians including John Cobb and Rosemary Ruether and Sally McFague using the prophetic voice to critique the underlying structures, religious and secular, that have played a role in the exploitation of natural resources. Another approach has been taken by Thomas Berry who, following the lead of Jesuit paleontologist Teilhard de Chardin, suggests that the book of nature itself provides sufficient revelation to warrant closer study, appreciation, and protection. Partnering with cosmologist Brian Swimme, Berry advocates the integration of religious and scientific narratives to cultivate a view that emphasizes intimacy and communion with the natural order rather than its utilitarian value for humans.

Yoga stands in the middle of this debate. As indicated at the start of this introduction, some notable scholars have argued in the manner of Lynn White, Jr., that the world-negating aspect of the religious philosophies of India, including Yoga, make Hinduism in most of its manifestations an unlikely candidate for inspiring ecological sensitivity or political advocacy leading to active protection or restoration of the wild. Several books have made a counter-argument to this position, particularly the work of political scientist O.P. Dwivedi[7] and physicist Vandana Shiva.[8] Several writers and scholars, included Ranchor Prime and David Haberman, have written eloquently of attempts to repair the environment of Vrindavan, a city sacred to Vaishnavas (worshippers of Krishna) and the Yamuna River which flows through it. Three primary anthologies explore both sides of this issue, with some essays suggesting that the Hindu religion has had an adverse ecological impact and others exploring the ways in which Hindu religious ideas and practices might help contribute to a solution: *Purifying the Earthly Body of God: Religion and Ecology in Hindu India*, edited by Lance E. Nelson (1998), *Ethical Perspectives on Environmental Issues in India*, edited by George A. James (1999), and *Hinduism and Ecology: The Intersection of Earth, Sky, and Water*, edited by Christopher Key Chapple and Mary Evelyn Tucker (2000).

The Yoga tradition in relation to ecology has been taken up in segments of various book chapters in the anthologies listed above. It also earned an individual entry in the *Encyclopedia of Religion, Ecology and Culture*, edited by Bron Taylor. On a more popular level, David Frawley's *Yoga and the Sacred Fire* explores Hindu and Yoga themes in light of nature romanticism. Henryk Skolimowski, Professor Emeritus of Philosophy and founder of the Eco-Philosophy Center at the University of Michigan, published many books and articles related to Yoga and ecology, including *Dharma, Ecology, and Wisdom in the Third Millenium* (1999) and *EcoYoga* (2003).

Building on the emerging field of Modern Yoga studies and her own activism, Laura Cornell, a contributor to this book, wrote a doctoral dissertation entitled "Green Yoga: A Collaborative Inquiry Among a Group of Kripalu Yoga Teachers" and established a newsletter that recently shifted to an on-line format. She also founded the Green Yoga Association which provides teacher training, conference, and a Green Studios Program. Popular magazines such as *Yoga Journal* (distribution 300,000) and *L.A. Yoga* (distribution 100,000) have devoted entire issues to the topic of Yoga and ecology. These activities indicate an enduring interest in exploring the contemporary relevance of Yoga in light of environmental ethics and eco-activism. With nearly 20 million regular practitioners of Yoga in the United States alone, this endeavor might bear fruitful results.

Yoga and Applied Environmental Ethics

This book begins with an examination of two sections of the *Atharva Veda*, one of the oldest texts of world literature, and concludes with an eye on the future. Though the Vedic peoples and the early progenitors of the Yoga tradition did not deal with such complex issues as species extinctions and global warming, they did observe nature and human nature closely. In the process, they developed cultural patterns (*saṃskāras*) that respect the realm of nature. Vedic texts and the Upaniṣads correlate the various realms of the cosmos to the body of a horse and to the human body, laying the foundation for a philosophy of interconnectedness between humans and non-human animals, and between the microcosm and the macrocosm. The Sāṃkhya tradition bestows upon the manifest world a feminine designation through its concept of *prakṛti*, perhaps drawing from the goddess hymns of the Vedas and anticipating the assertion of feminine power as Śakti in the later Tantric tradition. Though classical Yoga does not draw heavily from the Devi tradition,[9] and some have argued that Yoga de-emphasizes the physical world, the ethics of Yoga puts forth the notion that interaction within the world determines the quality of one's life and spirituality. Hence, three distinctive characteristics serve as a foundation for the yogic view of the universe: a bodily connection between the human realm and the universe, a lifting up of the feminine both in her particularity and as an organizing principle, and the importance of personal ethics.[10]

The medieval and modern phases of Yoga further explain the relevance of the manifest world in the quest for meaning. Tantra celebrates the realm of the senses, laying the foundation for creative engagement of the manifest for the purposes of both social uplift and personal liberation. For Abhinavagupta, the manifest world cannot be separated from the experience of a liberated consciousness. Though not discussed in this volume, the *Yogavāsiṣṭha* describes the five great elements of nature (*mahābhūta*s: earth, water, fire, air, space) as integral to the awakening into universal consciousness.

At the dawn of modernity in India, Mahatma Gandhi argued that the *Bhagavad Gītā*, which has been cited by some authors as antithetical to social justice values due to its emphasis on renunciation, in fact serves as a blueprint for both personal and societal deliverance. Arne Naess followed the Gandhian approach in his advocacy of Ecosophy T. In turn, this constructive synthesis has inspired thinkers and activists such as Henryk Skolomowski and Laura Cornell to boldly assert the relevance of Yoga to issues of ecological concern.

By presenting this collection of essays, the authors have engaged in a process and method of what may be deemed constructive theology. Using resources from ancient and medieval Hindu traditions, both as support and for purposes of dialogue and debate, this book develops ideas for an approach to environmental ethics informed and inspired by the Yoga tradition.

Endnotes

[1] Lance E. Nelson, ed. *Purifying the Earthly Body of God: Religion and Ecology in Hindu India* (Albany: State University of New York Press, 1998), p. 81.
[2] Lance E. Nelson in Christopher Key Chapple and Mary Evelyn Tucker, eds. *Hinduism and Ecology: The Intersection of Earth, Sky, and Water* (Cambridge: Harvard University Press, 2000), p. 151.
[3] Nelson, op.cit., pp. 95, 93.
[4] Ibid., p. 103.
[5] J. Baird Callicott. *Earth's Insights* (Berkeley: University of California Press, 1994), p. 48.
[6] James in Chapple, op.cit. 2000, p. 500
[7] The works of O. P. Dwivedi include *Environmental Crisis and Hindu Religion* (with B. N. Tiwari, New Delhi: Gitanjali Publishing House, 1987) and *India's Environmental Policies, Programmes, and Stewardship* (New York: St. Martin's Press, 1997).
[8] Vandana Shiva has published many books on related topics, including *Staying Alive: Women, Ecology, and Development* (London: Zed Books, 1988) and *Close to Home: Women Reconnect Ecology, Health, and Development Worldwide* (Philadelphia: New Society Publishers, 1994).
[9] One exception may be found in the choice of names for particular Yoga practices in the first section of the *Yoga Sūtra.* See "The Use of the of the Feminine Gender" in *Yoga and the Luminous: Patanjali's Spiritual Path to Freedom* (Albany: State University of New York Press, 2008), pp. 237–248.
[10] For further development of ecological descriptions, see my *Living Landscapes, Moving Beasts* (forthcoming).

Yoga and Nature: Vital Concentration in *Atharva Veda*

Suzanne Ironbiter

In which of its limbs does its fervor dwell? In which is its order set? Where its vow, where its trust? In what limb is its truth? (1)

In which limb dwells earth? In which dwells atmosphere? In which dwells sky? In which dwells heaven? (3)

In a person there join death and deathlessness, waters and channels. Tell me of Skambha, the one pole, what might it be? (15)

Who knows Spirit (Brahman) in a person knows the one most sought. (17a)

Within the world, in beings, on the back of the flowings, Spirit moves in fervor. In Him the powers of nature (devas) are as branches on a tree. (38)

The Pole Image

Several *Atharva Veda* hymns celebrate and inventory the mystical correspondences of external and internal nature, and of universal and personal order.[1] As in *Ṛg Veda* hymns interpreted psycho-philosophically,[2] the sequence of verses generates a dynamic series of focuses for intuitive concentration and meditative practice, partly through the double meanings, psychological and literal, of key words. In the chanting/reciting practice, the poet, priest, or participant enters the poem imaginatively as one would enter a mandala in an initiatory ritual journey. As Vedic *śruti*, the hymns are the singers' record of an illuminated divine reality revealed to them. The seer, in whose concentrated mind the focal sequence reveals itself, unscrolls for the ritual community the links that lead from image to image and stage to stage of the journey, moving from the outer layers to the unitive core, the yogic center connecting everything.

Among psycho-philosophical interpreters of the *Vedas,* V. S. Agrawala asks "what the Rishis intended to convey as a consistent formulation of their thoughts," and shows the interwoven spiritual and natural patterns in traditional esoteric interpretations of the *Vedas'* deity and ritual images.[3] Sri Aurobindo, Jan Gonda, Jeanine Miller, and Raimundo Panikkar also focus on the *Vedas'* mystical and psychological core.[4] Other writers (e.g., Pupul Jayakar, Madhu Khanna) point out the deep and primal ecological perceptions in the Vedic and autochthonous myths of India, perceptions akin to those of other indigenous and mystical traditions focused on experiencing spirit within nature.[5] Essays by K. L. Seshagiri Rao and T. S. Rukmani discuss traditional relationships between nature and dharmic, literary, and vital practices.[6]

This paper proposes to examine a proto-yogic example of spiritual union in the context of nature through a discussion of *Atharva Veda* 10.7, on *Skambha,* the Pole. Familiar from comparative religions, *Skambha*—pole stake, world pillar, central axis, world tree—is an archaic symbol of inter-relationship and orientation among the domains of experience: the upper, middle, and lower worlds (physical, psychic, and spiritual) and the sacred path for journeying between all aspects of reality.[7] The pole functions as a meditative center and repetitive refrain within the ritual motions and levels of the poem, of the poet's mind, of the dynamic world, and of historically subsequent cultural associations the poet/seer makes with it. There may be an implicit undeveloped association of the pole image with the Vedic *svaru* or *yupa*, the sacrificial post and its splinters, and with *Vanaspati*, the forest tree, lord of trees, source of ritual fuel and of timber for the household. Explicitly, however, as the poet's vision develops, he sees *Skambha* in cosmic human form (named as Vedic Puruṣa and Indra), and then esoterically as *jyeṣṭha Brahman*, *mahad yakṣaṃ*, and *guhya Prajāpati*, the hidden Lord of Life who knows *vetasaṃ hiraṇyaṃ,* the reed of gold within the *salila*, the flowings.

Doris Srinivasan, drawing on Louis Renou, gives a sacrificial interpretation of the poem: "the entire series of questions, which seek to associate different limbs of *Skambha* with different cosmic phenomena, bespeak of the dismemberment of a sacrificial victim" comparable to the dismemberment of Puruṣa in *RV* 10.90. Puruṣa, she says, represents a younger version of the Vedic Bull Asura, "a more ancient image of primeval matter" as the basis of the "cosmic parturition" that underlies the religious significance of divine multiple body parts in the *Ṛg Veda.* Following Renou, she cites the language of *AV* 10.7.20 as parallel to *RV* 10.90.9 in imagery of dismemberment.[8] On closer examination of those parallel stanzas, the *Skambha* hymn replaces the *RV*'s verb root *jan*/to give birth, with verb images for crafting associated with wood—*takṣ*/to hew, and *kaṣ*/to rub or scrape—consistent with the Pole/ World Tree image. Sri Aurobindo argues that modern Vedic scholars have tended to follow the medieval *RV* commentator Sayana's ritualistic sacrificial interpretations, having lost the ancient mystical core.

Pupul Jayakar observes that *Atharva Veda*, though compiled after *Ṛg Veda*, "contains very archaic elements and is possibly the earliest record of the beliefs, the imagery, the rituals and worships of the autochthonous peoples of India."[9] She sees, in some of the images and image sequences in the Indus seals, depictions of shapeshifting phenomena similar to those characteristic of indigenous spirit journeys worldwide. A well-known proto-Śiva yogi or *vrātya* image has such shape-shifted body parts as leaf-fronds for arms and an animal-mask head. Jayakar also shows a Harappan seal, dated third millennium BC, depicting a *Skambha*-like image, a "stone lingam, sacred pillar or trunk of tree with leaves sprouting from the living tree. Within the pillar is a standing male figure with arms stretched at his sides."[10] The figure has three radiation lines upward from the top of his head, lines such as one often sees worldwide in depictions of shamanic "solarization," representing, according to Joan Halifax, "the activation of the internal sun, the indwelling fire of life." [11]

In the current revision of the theory that the Vedic peoples invaded the Indus around 1500 BC, Indus visual artifacts and Vedic verbal artifacts are not from radically different cultures, but express a cultural continuum and may illuminate one another.[12] The mixture of archaic and Vedic images selected by the *Skambha* poet is particularly interesting to explore.

Reading *Atharva Veda* 10.7

The poem honors the natural environment through a serial focusing of the poet's attention on such categories as elements, measures of time, earth and celestial bodies, vertical and horizontal dimensions and directions, and 33 forces of nature (Rudras, Vasus, and Adityas). From the first verse onward, this outward attention is continuously added to and interwoven with attention to the internally grounded environment, the field of values, actions, questions, emotions, intentions: *tapas* (fervor), *ṛta* (order), *satya* (truth), *śrāddha* (faith), *vrata* (vows), *brahma* (prayers); desiring, longing, seeking, agreeing; measuring, purifying, arranging.

Grammatically, the poet's focusing sequence accentuates, in its verse-by-verse structure, the locative and ablative: in or from which limb or part, *aṅge*, of *Skambha*, the Pole, Sustainer or Support, are these external and internal aspects of life and experience to be found? In this way both external and internal features of life are drawn inward. Interwoven with the locative and ablative sequences are nominative interrogative refrains. Reminiscent of *RV* 10.121, where a single refrain concludes all but the last verse—"Who is this deity we will worship with our offerings?"[13]—the refrain variously asks: "Who is this Support? Who is this divine Being of so many parts, Whom the creator Prajapati fixed as the support of all worlds and all forms, Who is a who among many? How far does He enter or not enter all the vital parts of life He supports?" In the much shorter *Ṛg Vedic* hymn, the efficient purpose of the worship is to make a prayer-request to the all-pervading Lord of Creation; the overt focus remains on the natural environment and its blessings and treasures. Unlike in *AV* 10.7, the evolution of nature from Hiraṇyagarbha lacks a corresponding involution.

Searching out *Skambha*, the focusing vision of the poet begins, in verse 14, to see It as the support where the One *ṛṣi* and the ancient *ṛṣis* and Vedas abide. The *ṛṣis* and Vedas serve as internal and external human/divine intermediaries. Through their abiding, *Skambha* takes on the personal name and form of the Vedic Puruṣa. *Skambha*/Puruṣa is pictured as a kind of cosmic yogic body, a MahaPuruṣa whose *nāḍīs* are the vital lines of all things flowing and interrelating: "Who is this Puruṣa in whom are joined mortality and immortality, and in whose *nāḍīs* all flowings join, whose original *nāḍīs* hold the 4 quarters where *yajña*/worship/ritual crosses over?"(*AV* 10.7.15). Seeing that direct knowledge of *Skambha* comes also through knowledge of Puruṣa, Brahma, Prajapati, and Brahman, the poet connects the primordial *Skambha* image, and the insight contained within it, with these less archaic Vedic names,

names focusing prayer and concentration. In a reverse of the Puruṣa cosmogonic sacrifice of *RV* 10.90, he maps the macrocosmic and Vedic ritual spectrum onto specific parts of Puruṣa's body; it is an image of oneness and re-membering rather than dismembering and dispersion. In *Skambha*-Puruṣa, death and immortality, non-being and being, matter and spirit, are united (*samāhite*).

Grammatically, in contrast to *AV* 10.7's locatives, *RV* 10.90 features ablatives, showing how things come from the sacrifice; the poetic perspective of the body-parts symbolism is from the outside. The *RV* Puruṣa poem concludes, "The devas worshipped the sacrifice with sacrifice. These were the original *dharmas*/duties." The sacrificial theme is maintained to the end. The image is of spirit accepting partition by the natural world.

In *AV* 10.7, on the other hand, the re-membering process opens a deeper way into the Puruṣa image and the mystical core of the poem. Within the limbs or parts of Puruṣa's body, there is, the poet says, a hidden treasure which the 33 devas of the natural world—the Vasus, Rudras, and Ādityas—protect(*AV* 10.7.23). To know this hidden treasure, one must know the devas face to face in that place where they, knowing prayer, dwell in the ancient original sacred knowledge. "Great are those devas, abundantly sprung from what is not. People have called what is not the One limb of *Skambha*; bearing forth, He revolved that primeval limb in His limbs, and the 33 devas of the natural world disposed themselves in His limbs" (*AV* 10.7.25-27).

Knowing Brahma, the poet says, one knows those devas. "In the beginning, *Skambha* poured forth the golden treasure of Hirañyagarbha, the Golden Womb, in the world. On *Skambha* are the worlds, *tapas* (fervor), and *ṛta* (law) established. And I have directly seen you, Indra, established on *Skambha*" (*AV* 10.7.28-30). Perhaps referring to how early morning is the best time for prayer and visions, he adds, "Before dawn, one calls by name…"(*AV* 10.7.31). He pays homage to *Skambha* as *jyeṣṭa Brahman* (*AV* 10.7.32ff), most ancient, primeval focus of prayer upon which others such as the Vedic Indra become established, and in which others such as the devas of nature dispose themselves.

Drawing attention back to the flowing worlds of the mind and nature, and of the mind in nature, he asks, "Why doesn't the wind stand still, why doesn't the mind rest? Why, seeking truth, don't the waters stand still?" (*AV* 10.7.37). His answer is: "In the center of the world, behind the *salila*/flowing/waves/fluctuations, the great Spirit (*mahad yakṣaṃ*) is extended in *tapas*. In him all the devas rest like branches around a tree trunk. Truly the hidden Prajapati, Lord of Life knows the reed of gold set in the *salila*/ flowing/ waves/ fluctuations" (*AV* 10.7.38&41). The esoteric image of the reed of gold—*vetasaṃ hiraṇyaṃ*—suggests, as in *RV* 4.58.5, the visionary "solarization" core mirrored in natural, mystical, and ritual Vedic images for focusing thought on light: dawn, sun, fire, flows of *ghṛta*, and the ritual and symbolic fire hidden in wood.

The poem, reflecting an image from *Ṛg Veda* 10.130, concludes with a vision of cosmic, meditative, and ritual weaving in which two young women, light and dark, day and night, heaven and earth, repeatedly lay warp and woof toward the directions, while *Pumān,* the cosmic man, suggestive of Puruṣa, inweaves and extends the web, using *sāman* (songs) made as shuttles (*AV* 10.7.42–44). In *RV* 10.130, the poet sees with the eye of his mind how *Pumān* stretches the warp and woof of the primal ritual, and makes the models for the *yajña*'s song meters. Thus the *Skambha* poem concludes with a vision of poesis and cosmos projected outward from the light within the ever-dynamic internal and external weave.

In summary, the poet envisions *Skambha* as alive with ritual values, as the resting and connecting place of the devas of nature, and as containing an all-illuminating vision. Agni, rising, shines with desire toward the mystical Pole; waters flow with desire toward the Pole. *Skambha*'s mouth is Brahma, His tongue is *madhukaṣa*/honey-whip, His bosom is *Virāj* (*AV* 10.7.19). The *Devas* of nature, with hands, feet, word, ear, and eye, offer tribute (*balim prayacchanti*, from *yam*, to stretch forth) to the unmeasurable with the measurable (*AV* 10.7.39).

Conclusion: The Hymn and the Web of Life

Eliade suggests that poetic arts originated in primordial shamanic techniques for healing, defending, and maintaining "the psychic integrity of the community."[14] In indigenous views worldwide, community includes "all my relations," that is every being's interconnection with every other being. Concentration, ritual, and song maintain awareness of the entire web of natural, social, psychological, and spiritual reality. The poet of the *Skambha* hymn, with his images of pole, tree, indwelling light, and web, is in a direct lineage with this view. His poem does not feature partitional imagery of symbolic or real animal sacrifice. It speaks of the connective need to know, face to face, the primordial devas of the natural world, disposed in *Skambha*'s limbs, and their ancient original sacred prayer knowledge. The poet and the poem stretch forth with the measurable to the unmeasurable, which then illuminates the measurable and maintains the inner and outer harmonies of life, causing us to remember the vital cycles of exchange.

Endnotes

---

1 e.g., *AV* 2.1, 5.1, 10.1, 10.8, 11.8, 15.8, 19.6.

2 Jeanine Miller, *The Vedas* (London: Rider & Company, 1974), xxii.

---

[3] V. S. Agrawala, *Vedic Lectures* (Varanasi: Prithivi Prakashan, 1982), p. vii. (Lectures from 1960).
[4] Sri Aurobindo, *The Secret of the Veda* (Pondicherry: Sri Aurobindo Ashram, 2004). (Writings from 1914–1920).
J. Gonda, *The Vision of the Vedic Poet* (New Delhi: Munshiram Manoharlal, 1984).
J. Gonda and Jeanine Miller, "The Heart of Rgvedic Religion," *The Essence of Yoga.* Ed. Georg Feuerstein & Jeanine Miller(Rochester, VT: Inner Traditions International, 1998) pp. 121–151. (1st edition 1971).
Raimoundo Panikkar, *Mantramanjari: The Vedic Experience* (Berkeley: University of California Press, 1977).
[5] Pupul Jayakar, *The Earthen Drum* (New Delhi: National Museum, 1980).
Madhu Khanna, "Nature as Feminine." *Man in Nature,* Vol. 5., Ed. Baidyanath Saraswati. (New Delhi: Indira Gandhi National Centre for the Arts, 1995).
[6] K. L. Seshagiri Rao, "The Five Great Elements: An Ecological Perspective." *Hinduism and Ecology.* Ed. Christopher Key Chapple & Mary Evelyn Tucker (Cambridge: Harvard University Press, 2000), pp. 23–38. T. S. Rukmani. "Literary Foundations for an Ecological Aesthetic." Ibid., pp. 101–125.
[7] Mircea Eliade, *Shamanism* (Princeton: Princeton University Press, 1964), ch. 8.
[8] Doris Srinivasan, "Religious Significance of Divine Multiple Body Parts in the *Atharva Veda,*" *Numen.* 25.3 (1978), p. 212.
[9] Ibid., p. 36.
[10] Ibid., p. 41.
[11] Joan Halifax, *Shaman* (London: Thames & Hudson, 1982), pp. 90–91.
[12] cf. Sri Aurobindo, op.cit.
Georg Feuerstein, Subhash Kak & David Frawley, *In Search of the Cradle of Civilization* (Wheaton, IL: Quest Books, 2001).

[13] *RV* and *AV* translations are my own. Texts consulted include: *Rig Veda Samhita*, Ed. R. L. Kashyap & S. Sadagopan (Bangalore: Sri Aurobindo Kapali Sastry Institute of Vedic Culture, 1998).
The Hymns of the Rgveda, Trans Ralph T. H. Griffith (Delhi: Motilal Banarsidass, 1973).
The Atharvaveda, Ed. Devi Chand (Delhi: Munshiram Manoharlal, 2002).
Atharva-Veda-Samhita, Trans. William Dwight Whitney (Delhi: Motilal Banarsidass, 2001).
Hymns from the Vedas, Ed. and trans.Abinash Chandra Bose (Bombay: Asia Publishing House, 1966).
Panikkar, op.cit.
[14] Eliade, p. 508.

Pṛthivī Sūkta: Earth Verses

Christopher Key Chapple

Suzanne Ironbiter has introduced us to some key themes from Vedic literature, especially the "mystical correspondences of external and internal nature" and "deep and primal ecological perceptions," particularly in the image of Skambha the Sustainer in the *Atharva Veda*. This text, known for its medical lore, also includes one section in praise of the earth: the *Pṛthivī Sūkta*, or stanzas pertaining to the Earth. It comprises the first 63 verses of the twelfth book of this important work. Because of its extended length and its thematic consistency, it provides a ready window into the earth-consciousness of Vedic India. Lynn White, Jr., found fault with the early biblical material, suggesting that it paved the way for the rape of nature. In reading the *Pṛthivī Sūkta*, one gets somewhat an opposite impression. The author does not urge the reader to ravage the earth, does not provide advice on how to obtain food, nor does he or she laud the human as superior to the earth in any way. Instead, the *Pṛthivī Sūkta* describes the abundance of the earth and asks that she protect humans. In turn the text urges humans to protect the earth.

The *Pṛthivī Sūkta*: Praise, Petition, and Admonition[1]

The *Pṛthivī Sūkta* includes various themes and approaches to the earth. Many of the verses praise the earth, and, generally, after giving praise, the writer of the hymn asks for protection and support from the earth. The *Pṛthivī Sūkta* emphasizes the relationship between fragrance and the earth, as well its connection with water and fire. It links the protection of the earth to the god Indra, and echoes the statement in the *Ṛg Veda* regarding the creation of the earth by the three strides of Vishnu. It acknowledges that all works and all religious endeavors rely upon the earth, and that medicines arise from the earth. It includes two stanzas that seem to speak directly to our current environmental situation, one referring to water pollution (30) and the other to the development of roads (47). After discussing each of these themes, some reflections will follow regarding the applicability of these ancient musings to the development of an ecological ethic for contemporary India.

The praise verses of the *Pṛthivī Sūkta* provide rich imagery of the earth. The author lists here many attributes, noting that she is "adorned with many hills, plains, and slopes" (2). The physical landscape is described in detail, with the author proclaiming that "Upon the earth lie the oceans, many rivers, and other bodies of water" (3), that "On her body food is grown everywhere and on her the farmer toils" (4), and that "The earth is the home of cows, horses, and of birds" (5). The text continues with additional details on the grandeur of the earth: "Sacred are your hills, snowy mountains, and deep forests" (11). Verse 56 praises a variety of earth's places:

> Irrespective of the place and region where we are,
> whether in a rural area, in the woods,
> in the battleground, or in a public place,
> may we always sing your praises.

The earth is seen in a couple verses as the "wish fulfilling cow (*kāmadhenu*)" (45, 61) and lauded for her expansiveness and generosity: "You are borderless, you are the world-mother of all things, you are the provider of all things in life" (61). Acknowledgement can be found of human reliance on the earth in several verses, such as "My mother is this earth, and I am her son" (12), "You are the world for us and we are your children" (16), with reference to her support for the many races and nations found on earth (11) and "The five races of human beings [that] live here" (42).

While these verses typologize the various functions and forms of earth, the *Pṛthivī Sūkta* also carries a clearly anthropocentric message. In verse after verse, the author petitions the earth to render protection and succour to human beings: "May the world Mother provide us with a wide and limitless domain for our livelihood" (1), "may she spread prosperity for us all around" (2), "May the Earth confer on us all riches" (3), "May that Earth replenish us in plenty with cattle and food" (4), "May that Earth protect us, grant us prosperity, and bestow upon us vigor" (5), "May you give us wealth and good fortune!" (6), "May She make our nation strong, powerful, and studded with splendour" (8), and so forth. Nearly every verse ends with a request for enhanced well being. An extended set of verses also discusses interpersonal relationships, with the author asking that the Earth spare one from hatred and gird one in battle against enemies (18, 24, 25, 32, 37, 41). Additionally, several verses ask for protection for harm and danger to be found in the natural world:

> Keep away from us venomous reptiles such as snakes and scorpions which cause thirst when they sting; keep away those poisonous insects which cause fever, and let all those terrible crawling creatures which are born in the rainy season keep away from us. (46)
> Keep all menacing animals away for harming us, such as the lion, the tiger, the wolf, the jackal. (49)

The *Pṛthivī Sūkta* does not only extol nature but sees a need to profit from nature and to be wary of nature's dark side.

In a metaphysical sense, the text reflects the Sāṃkhya philosophy of the great elements (*mahābhūta*s) and their relationship to the subtle elements (*tanmātra*s) and the sense organs (*buddhīndriya*s). The *Sāṃkhya Kārikā* aligns the earth, also referred to as *pṛthivī*, with the sense of smell and fragrance, a theme found in several verses of the *Pṛthivī Sūkta*:

> O Mother Earth!
> Instill in me with abundance that fragrance which emanates from you and from you herbs and other vegetation, as well as waters.
> This fragrance is sought by all celestial beings. (23)
> O Mother Earth!
> May that perfume come to us in abundance, the perfume that is in the lotus, the fragrance worn by gods when the sun marries the dawn. (24)
> O Mother Earth!
> The fragrance that you have granted to men and women and which is also present in horses, deer, and elephants, shines like the radiance in maidens.
> May that radiance come to us. (25)
> Bless us with.. fragrance.
> Grant us peace, tranquillity, fragrant air and other worldly riches. (59)

Several other verses also refer to the great elements, particularly fire (*agni*) and water (*āp*). Alluding to earlier texts of the *Ṛg Veda* and the *Upaniṣads*, the *Pṛthivī Sūkta* asserts that the earth arose from the water of the ocean (8). For a series of verses it extols the power of fire, claiming that fire energizes the earths's herbs and medicinal plants, as well as the clouds that give rain to the earth (19–21).

The gods are spoken of in relationship to the earth. Agni, the aniconic god of fire, has been mentioned above. Two anthropomorphized gods find special mention: Indra and Vishnu. In the *Ṛg Veda*, Indra conquers the dragon (*Vṛtra*) of drought, frees the monsoon, and allows structure to emerge on earth (*RV* X:125). Verse 37 echoes this episode, which states that the earth chose "Indra as her companion rather than Vṛtra." The text refers to Indra as the consort of the earth (6) and states that she is "protected by the great strength of Indra" (18). It also states that she provides the place where "Indra is invoked to drink Soma," the substance used for religious inspiration (38). Verse 10 alludes to the famous stanza in the *Ṛg Veda* that describes how Vishnu created the earth in three great strides.

The earth supports religious ritual and sacrifice and ensures social stability. She yields the medicine used in healing practices (17, 19, 20, 62). She houses the "sacred universal fire" (6), the "ever-vigilant and all-caring gods" (7), and serves as the site for sacrifices:

> Events for the welfare of all are consecrated by performing sacrifices on this Earth.
> Good and virtuous people assemble here to perform such functions.
> Strong sacrificial posts are erected here for making offerings.
> Here is where spirituality gets imparted. (13)
> Oblations and sacrifices are duly performed for Mother Earth. (22)
> Places for oblations can be found on this Earth.
> Here can be found the poles for the sacrifice.
> This is where the sacrificial post is situated.
> This is where Brahmans well-versed in the Vedas recite hymns. (38)
> It is upon this Earth where the Seven Sages, the creators of worlds, performed sacrifices and austerities,
> chanted hymns, and carried out sacred rites. (39)

In both an expansive, cosmic sense, the Earth contains all things. The Earth also provides the specific ground upon which human beings may acknowledge their dependence on the Earth through religious activities.

The *Pṛthivī Sūkta* also proclaims that, even in light of her magnitude and magnificence, humans need to do their best to not harm or injure the Earth. Along with petitioning the Earth for her blessings, the author of the text puts forth an admonition that the Earth must be guarded from human interference in the following passages: "May no person oppress her" (2), "May we never harm your vital parts" (35), "Please do not become outraged by our destructive tendencies" (45), and "May I have the strength to subjugate those who poison our Mother Earth" (54). Toward the end of the text, the author's vehemence increases: "If anyone tries to harm you, destroy them with the same ease as when a horse shakes off the dust on it" (57) and "May I have the ability of swiftness and strength so that I am able to vanquish those who exploit you" (58). These verses seem to anticipate a misuse of the earth and perhaps reflect some of the rapid changes inflicted by human interference during the settling of the Gangetic Plains.

Two verses in particular indicate that human impact was being felt upon the earth. In verse 30, a specific warning is given in regard to water pollution.

> O Mother Earth!
> May our bodies enjoy only the clean water.
> May you keep away from us that which is polluted
> and may we do only the good deeds.

Though pollution can arise from many sources, this verse indicates an awareness that unclean water can lead to disease and death.

Verse 47 acknowledges that human expansion and road building have altered the face of the earth. In the following verse, the author states that this has been freely given by the earth, but asks for protection on these roads:

> O Mother Earth!
> You have given people many roads where both the chariots move and bullock-carts with grains ply,
> where both the virtuous and wicked people travel.
> Protect these networks of transportation from robbers and thugs.
> May we be victorious and may we receive the auspicious and benevolent things in life.

The first verse asks for wholesome resources; the second verses acknowledges that trouble can arise where human beings congregate. If we extend the metaphor from this juxtaposition, human beings need clean resources to flourish. If these become endangered, then stability can founder.

In an expression of proto-environmentalism, and in a sense responding to the problem articulated above, verse 27 makes an appeal for venerating the Earth, both for her sake, and for the sake of human welfare:

> We venerate Mother Earth,
> the sustainer and preserver of forests, vegetation,
> and all things that are held together firmly.
> She is the source of a stable environment.

Without remembering Mother Earth, forests disappear, water becomes foul, and the air becomes unbreatheable.

This foundational Vedic text colourfully illustrates the importance of the earth element. She is depicted as the matrix, the origin of all material particularity. As an object of worship and veneration, she becomes the emblem for the relationship between humans and the earth.

From Vedas to Yoga

This ancient hymn establishes and affirms the importance of the earth as mother in Indian civilization. In the philosophical traditions that followed the Vedas and the Upaniṣads, all things related to the earth and the manifest world were gathered together under a single ontological principle: *prakṛti*, the matrix of creativity. Working through three qualities or sets of characteristics (*guṇas*), known as heaviness (*tamas*), energetic movement (*rajas*), and illumination (*sattva*), *prakṛti* brings forth the earth as well as the other elements, the human body, the mind, and the emotions. Her task and hence the task of the earth itself is to provide experience for consciousness. Because this consciousness (*puruṣa* or *ātman*) stands apart from *prakṛti*, some have argued that her work carries little meaning or spiritual significance. However, the essays that follow take the position that *prakṛti* holds the only pathway to meaning. Without a world, there can be no experience. Without experience there can be no awareness and no liberation.

The Skambha and Pṛthivī Sukta sections of the *Atharva Veda* lay the foundation for taking the material world seriously. By venerating the magnificence of creation and by establishing correlations between the microphase human realm and the macrophase cosmic realm, the *Atharva Veda* anticipates the insights and practices of later meditative traditions. By extolling and recognizing the beauty of nature, a sense of connectivity may be experienced that bears relevance to the development of Yoga in relation to environmental ethics.

Endnotes

[1] The translations from the *Pṛthivī Sūkta* have been rendered by O. P. Dwivedi and edited by the author as part of a forthcoming book project. Earlier translations of the text include William Dwight Whitney, tr., *Atharva-Veda Saṃhita*, Volume II (Delhi, Motilal Banarsidass, 1962), pp. 660–672 and Shrinivas S. Sohoni, *Hymn to the Earth: The Prithivi Sukta* (Delhi: Sterling Publishers, no date).

The Integration of Spirit (*Puruṣa*) and Matter (*Prakṛti*) in the *Yoga Sūtra*

Ian Whicher

Introduction

This paper centers on the thought of Patañjali (ca second–third century CE), the great exponent of the authoritative classical Yoga school (*darśana*) of Hinduism and the reputed author of the *Yoga Sūtra*. I will argue that Patañjali's philosophical perspective has, far too often, been looked upon as excessively "spiritual" or isolationistic to the point of being a world-denying philosophy, indifferent to moral endeavor, neglecting the world of nature and culture, and overlooking the highest potentials for human reality, vitality, and creativity. Contrary to the arguments presented by many scholars, which associate Patañjali's Yoga exclusively with extreme asceticism, mortification, denial, and the renunciation and abandonment of "material existence" (*prakṛti*) in favor of an elevated and isolated "spiritual state" (*puruṣa*) or disembodied state of spiritual liberation, I suggest that Patañjali's Yoga can be seen as a responsible engagement, in various ways, of "spirit" (*puruṣa* = intrinsic identity as Self, pure consciousness) and "matter" (*prakṛti* = the source of psychophysical being, which includes mind, body, nature) resulting in a highly developed, transformed, and participatory human nature and identity, an integrated and embodied state of liberated selfhood (*jīvanmukti*).

The interpretation of Patañjali's Yoga Darśana presented in this paper—which walks the line between an historical and hermeneutic-praxis (some might say theological or "systematic") orientation—counters the radically dualistic, isolationistic, and ontologically oriented interpretations of Yoga[1] presented by many scholars and suggests an open-ended, epistemologically oriented hermeneutic which, I maintain, is more appropriate for arriving at a genuine assessment of Patañjali's system.

It is often said that, like classical Sāṃkhya, Patañjali's Yoga is a dualistic system, understood in terms of *puruṣa* and *prakṛti*. Yet, I submit, Yoga scholarship has not clarified what "dualistic" means or why Yoga had to be "dualistic." Even in avowedly non-dualistic systems of thought such as Advaita Vedānta we can find numerous examples of basically dualistic modes of description and explanation.[2]

Elsewhere[3] I have suggested the possibility of Patañjali having asserted a provisional, descriptive, and "practical" metaphysics, i.e., in the YS[*] the metaphysical schematic is abstracted from yogic experience, whereas in classical Sāṃkhya, as set out in Īśvara Kṛṣṇa's *Sāṃkhyakārikā*, "experiences" are fitted into a metaphysical structure. This approach would allow the YS to be interpreted along more open-ended, epistemologically oriented lines without being held captive by the radical, dualistic metaphysics of Sāṃkhya. Despite intentions to render the experiential dimension of Yoga, purged as far as possible from abstract metaphysical knowledge, many scholars have fallen prey to reading the YS from the most abstract level of the dualism of *puruṣa* and *prakṛti* down to an

---

* Abbreviations are as follows:

BG	Bhagavadgītā
RM	Rāja-Mārtaṇḍa of Bhoja Rāja (ca eleventh century CE)
SK	Sāṃkhya-Kārikā of Īśvara Kṛṣṇa (ca fourth-fifth century CE)
TV	Tattva-Vaiśāradī of Vācaspati Miśra (ca ninth century CE)
YB	Yoga- Bhāṣya of Vyāsa (ca fifth-sixth century CE)
YS	Yoga-Sūtra of Patañjali (ca second-third century CE)
YSS	Yoga-Sāra-Saṃgraha of Vijñāna Bhikṣu (ca sixteenth century CE)
YV	Yoga-Vārttika of Vijñāna Bhikṣu

understanding of the practices advocated. Then they proceed to impute an experiential foundation to the whole scheme informed not from mystical insight or yogic experience, but from the effort to form a consistent (dualistic) world-view, a view that culminates in a radical dualistic finality[4] or closure.

Patañjali's philosophy is not based upon mere theoretical or speculative knowledge. It elicits a practical, pragmatic, experiential/perceptual (not merely inferential/theoretical) approach that Patañjali deems essential in order to deal effectively with our total human situation and provide real freedom, not just a theory of liberation or a metaphysical explanation of life. Yoga is not content with knowledge (*jñāna*) perceived as a state that abstracts away from the world removing us from our human embodiment and activity in the world. Rather, Yoga emphasizes knowledge in the integrity of being and action and as serving the integration of the "person" as a "whole." Edgerton concluded in a study dedicated to the meaning of Yoga that: "... Yoga is not a 'system' of belief or of metaphysics. It is always a way, a method of getting something, usually salvation... ."[5] But this does not say enough, does not fully take into account what might be called the integrity of Patañjali's Yoga. Yoga derives its real strength and value through an integration of theory and practice.[6]

Cessation (Nirodha) and the 'Return to the Source' (Pratiprasava): Transformation or Elimination/Negation of the Mind?

In Patañjali's central definition of Yoga, Yoga is defined as "the cessation (*nirodha*) of [the misidentification with] the modifications (*vṛtti*) of the mind (*citta*)".[7] What kind of "cessation" we must ask is Patañjali actually referring to in his classical definition of Yoga? What does the process of cessation actually entail for the yogin ethically, epistemologically, ontologically, psychologically, and so on? I have elsewhere suggested[8] that *nirodha* denotes an epistemological emphasis and refers to the transformation of self-understanding brought about through the purification and illumination of consciousness; *nirodha* is not (for the yogin) the ontological cessation of *prakṛti* (i.e., the mind and *vṛttis*). Seen here, *nirodha* thus is not, as is often explained, an inward movement that annihilates or suppresses *vṛttis*, thoughts, intentions, or ideas (*pratyaya*), nor is it the nonexistence or absence of *vṛtti*; rather, *nirodha* involves a progressive unfoldment of perception (*yogi-pratyakṣa*) that eventually reveals our true identity as *puruṣa*. It is the state of affliction (*kleśa*) evidenced in the mind and not the mind itself that is at issue. *Cittavṛtti* does not stand for all modifications or mental processes (cognitive, affective, emotive), but is the very seed (*bīja*) mechanism of the misidentification with *prakṛti* from which all other *vṛttis* and thoughts arise and are (mis)appropriated or self-referenced in the state of ignorance (*avidyā*), that is, the unenlightened state of mind. Spiritual ignorance gives rise to a malfunctioning or misalignment of *vṛtti* with consciousness that in Yoga can be corrected thereby allowing for a proper alignment or "right" functioning of *vṛtti*.[9] It is the *cittavṛtti* as our confused and mistaken identity, not our *vṛttis*, thoughts, and experiences in total that must be brought to a state of definitive cessation. To be sure, there is a temporary suspension of all the mental processes as well as any identification with an object (i.e., in *asaṃprajñāta-samādhi*, this being for the final purification of the mind[10]), but it would be misleading to conclude that higher *samādhi* results in a permanent or definitive cessation of the *vṛttis* in total thereby predisposing the yogin who has attained purity of mind to exist in an incapacitated, isolated, or mindless state and therefore of being incapable of living a balanced, useful, and productive life in various ways.

From the perspective of the discerning yogin (*vivekin*), human identity *contained* within the domain of the three *guṇas* of *prakṛti* (i.e., *sattva*, *rajas*, and *tamas*) amounts to nothing more than sorrow and dissatisfaction (*duḥkha*).[11] The declared goal of classical Yoga, as with Sàṃkhya and Buddhism, is to overcome all dissatisfaction (*duḥkha*, YS II.16) by bringing about an inverse movement or counter-flow (*pratiprasava*)[12] understood as a "return to the origin"[13] or "process-of-involution"[14] of the *guṇas*, a kind of reabsorption into the transcendent purity of being itself. What does this "process-of-involution"—variously referred to as "return to the origin," "dissolution into the source"[15] or "withdrawal from manifestation"—actually mean? Is it a definitive ending to the perceived world of the yogin comprised of change and transformation, forms and phenomena? Ontologically conceived, *prasava* signifies the "flowing forth" of the primary constituents or qualities of *prakṛti* into the multiple forms of the universe in all its dimensions, i.e., all the processes of manifestation and actualization or "creation" (*sarga*, *prasarga*). *Pratiprasava* on the other hand denotes the process of "dissolution into the source" or "withdrawal from manifestation" of those forms relative to the personal, microcosmic level of the yogin who is about to attain freedom (*apavarga*).

Does a "return to the origin" culminate in a state of freedom in which one is stripped of all human identity and void of any association with the world including one's practical livelihood? The ontological emphasis usually given to the meaning of *pratiprasava*—implying for the yogin a literal dissolution of *prakṛti's* manifestation—would seem to support a view, one which is prominent in Yoga scholarship, of spiritual liberation denoting an existence wholly transcendent (and therefore stripped or deprived) of all manifestation including the human relational sphere. Is this the kind of spiritually emancipated state that Patañjali had in mind (pun included)? In YS II.3–17 (which set the stage for the remainder of the chapter on yogic means or *sādhana*), Patañjali describes *prakṛti*, the "seeable" (including our personhood), in the context of the various afflictions (*kleśas*) that give rise to an afflicted and mistaken identity of self. Afflicted identity is constructed out of and held captive by the root affliction of ignorance (*avidyā*) and its various forms of karmic bondage. Yet, despite the clear association of *prakṛti* with the bondage of ignorance (*avidyā*), there are no real grounds for purporting that *prakṛti* herself is to be equated with or subsumed under the afflictions. To equate *prakṛti* with affliction itself implies that as a product of spiritual ignorance, *prakṛti*, along with the afflictions, is conceived as a reality that the yogin should ultimately abandon, condemn, avoid or discard completely. Patañjali leaves much room for understanding "dissolution" or "return to the source" with an epistemological emphasis thereby allowing the whole system of the Yoga Darśana to be interpreted along more open-ended lines. In other words, what actually "dissolves" or is ended in Yoga is the yogin's misidentification with *prakṛti*, a mistaken identity of self that—contrary to authentic identity, namely *puruṣa*—can be nothing more than a product of the three *guṇas* under the influence of spiritual ignorance. Understood as such, *pratiprasava* need not denote the definitive ontological dissolution of manifest *prakṛti* for the yogin, but rather refers to the process of "subtilization" or sattvification of consciousness so necessary for the uprooting of misidentification—the incorrect world-view born of *avidyā*—or incapacity of the yogin to "see" from the yogic perspective of the seer (*draṣṭṛ*), our authentic identity as *puruṣa*.

The discerning yogin sees (YS II.15) that this guṇic world or cycle of saṃsāric identity is in itself dissatisfaction (*duḥkha*). But we must ask, what exactly is the problem being addressed in Yoga? What is at issue in Yoga philosophy? Is our ontological status as a human being involved in day to day existence forever in doubt, in fact in need of being negated, dissolved in order for authentic identity (*puruṣa*), immortal consciousness, finally to dawn? Having overcome all ignorance, is it then possible for a human being to live in the world and no longer be in conflict with oneself and the world?

Can the *guṇas* cease to function in a state of ignorance and conflict in the mind? Must the guṇic constitution of the human mind and the whole of prakṛtic existence disappear, dissolve for the yogin? Can the ways of spiritual ignorance be replaced by an aware, conscious, nonafflicted identity and activity that transcend the conflict and confusion of ordinary, saṃsāric life? Can we live, according to Patañjali's Yoga, an embodied state of freedom?

"Aloneness" (Kaivalya): Implications for an Embodied Freedom

In the classical traditions of Sāṃkhya and Yoga, *kaivalya*, meaning "aloneness,"[16] is generally understood to be the state of the unconditional existence of *puruṣa*. In the YS, *kaivalya* can refer more precisely to the "aloneness of seeing" (*dṛśeḥ kaivalyam*) which, as Patañjali states, follows from the disappearance of ignorance (*avidyā*) and its creation of *samyoga*[17]—the conjunction of the seer (*puruṣa*) and the seeable (i.e. *citta*, *guṇas*)—explained by Vyāsa as a mental superimposition (*adhyāropa*, YB II.18) . "Aloneness" thus can be construed as *puruṣa*'s innate capacity for pure, unbroken, non-attached seeing/perceiving, observing or "knowing" of the content of the mind (*citta*).[18] In an alternative definition, Patañjali explains *kaivalya* as the "return to the origin" (*pratiprasava*) of the *guṇas*, which have lost all soteriological purpose for the puruṣa that has, as it were, recovered its transcendent autonomy.[19] This *sūtra* (YS IV.34) also classifies *kaivalya* as the establishment in "own form/nature" (*svarūpa*), and the power of higher awareness (*citiśakti*).[20] Although the seer's (*draṣṭṛ/puruṣa*) capacity for "seeing" is an unchanging yet dynamic power of consciousness that should not be truncated in any way, nevertheless our karmically distorted or skewed perceptions vitiate against the natural fullness of "seeing." Patañjali defines spiritual ignorance (*avidyā*), the root affliction, as: "seeing the noneternal as eternal, the impure as pure, dissatisfaction as happiness, and the nonself as self" (YS II.5). Having removed the "failure-to-see" (*adarśana*), the soteriological purpose of the *guṇas* in the saṃsāric condition of the mind is fulfilled; the mind is relieved of its former role of being a vehicle for *avidyā*, the locus of egoity and misidentification, and the realization of pure seeing—the nature of the seer alone—takes place.

According to yet another *sūtra* (YS III.55), we are told that *kaivalya* is established when the *sattva* of consciousness has reached a state of purity analogous to that of the *puruṣa*.[21] Through the process of subtilization or "return to the origin" (*pratiprasava*) in the *sāttva*, the transformation (*pariṇāma*) of the mind (*citta*) takes place at the deepest level bringing about a radical change in perspective: the former impure, fabricated states constituting a fractured identity of self are dissolved resulting in the complete purification of mind. Through knowledge (in *saṃprajñāta-samādhi*) and its transcendence (in *asaṃprajñāta-samādhi*) self-identity overcomes its lack of intrinsic grounding, a lack sustained and exacerbated by the web of afflictions in the form of attachment, aversion, and the compulsive clinging to life based on the fear of extinction. The yogin is no longer dependent on liberating knowledge (mind-*sattva*),[22] is no longer attached to *vṛtti* as a basis for self-identity. Cessation, it must be emphasized, does not mark a definitive disappearance of the *guṇas* from *puruṣa*'s view.[23] For the liberated yogin, the *guṇas* cease to exist in the form of *avidyā* and its *saṃskāras*, *vṛttis*, and false or fixed ideas (*pratyaya*) of selfhood that formerly veiled true identity. The changing gunic modes cannot alter the yogin's now purified and firmly established consciousness. The mind has been liberated from the egocentric world of attachment to things prakrtic. Now the yogin's identity (as *puruṣa*), disassociated from ignorance, is untouched, unaffected by qualities of mind,[24] uninfluenced by the *vṛttis* constituted of the three *guṇas*. The mind and *puruṣa* attain to a sameness of purity (YS III.55), of harmony, balance, evenness, and a workability together: the mind appearing in the nature of *puruṣa*.[25]

Kaivalya, I suggest, in no way destroys or negates the personality of the yogin, but is an unconditional state in which all the obstacles or distractions preventing an immanent and purified relationship or engagement of person with nature and spirit (*puruṣa*) have been removed. The mind, which previously functioned under the sway of ignorance coloring and blocking our perception of authentic identity, has now become purified and no longer operates as a locus of misidentification, confusion, and dissatisfaction (*duḥkha*). *Sattva*, the finest quality (*guṇa*) of the mind, has the capacity to be perfectly lucid/transparent, like a dust-free mirror in which the light of *puruṣa* is clearly reflected and the discriminative discernment (*vivekakhyāti*)[26] between *puruṣa* and the *sattva* of the mind (as the finest nature of the seeable) can take place.[27]

The crucial (ontological) point to be made here is that in the "aloneness" of *kaivalya prakṛti* ceases to perform an obstructing role. In effect, *prakṛti* herself has become purified, illuminated, and liberated[28] from *avidyā's* grip including the misconceptions, misappropriations, and misguided relations implicit within a world of afflicted identity. The mind has been transformed, liberated from the egocentric world of attachment, its former afflicted nature abolished; and self-identity left alone in its "own form" or true nature as *puruṣa* is never again confused with all the relational acts, intentions, and volitions of empirical existence. There being no power of misidentification remaining in *nirbīja-samādhi*,[29] the mind ceases to operate within the context of the afflictions, karmic accumulations, and consequent cycles of *saṃsāra* implying a mistaken identity of selfhood subject to birth and death.

The *Yoga Sūtra* has often been regarded as calling for the severance of *puruṣa* from *prakṛti*; concepts such as liberation, cessation, detachment/dispassion, and so forth have been interpreted in an explicitly negative light. Max Müller, citing Bhoja Rāja's commentary[30] (eleventh century CE), refers to Yoga as "separation" (*viyoga*).[31] More recently, numerous other scholars[32] have endorsed this interpretation, that is, the absolute separateness of *puruṣa* and *prakṛti*. In asserting the absolute separation of *puruṣa* and *prakṛti*, scholars and non-*scholars* alike have tended to disregard the possibility for other (fresh) hermeneutical options, and this radical, dualistic metaphysical closure of sorts surrounding the nature and meaning of Patañjali's Yoga has proved detrimental to a fuller understanding of the Yoga Darśana by continuing a tradition based on an isolationistic, one-sided reading (or perhaps misreading) of the YS and Vyāsa's commentary. Accordingly, the absolute separation of *puruṣa* and *prakṛti* can only be interpreted as a disembodied state implying death to the physical body. To dislodge the sage from bodily existence is to undermine the integrity of the pedagogical context that lends so much credibility or "weight" to the Yoga system. I am not here implying a simple idealization of Yoga pedagogy thereby overlooking the need to incorporate a healthy critical approach to the guru-disciple dynamic. Rather, I am suggesting that it need not be assumed that, in Yoga, liberation coincides with physical death.[33] This would only allow for a soteriological end state of "disembodied liberation" (*videhamukti*). What is involved in Yoga is the death of the atomistic, egoic identity, the dissolution of the karmic web of *saṃsāra* that generates notions of one being a subject trapped in the prakṛtic constitution of a particular body-mind.

Not being content with mere theoretical knowledge, Yoga is committed to a practical way of life. To this end, Patañjali included in his presentation of Yoga an outline of the "eight-limbed" path (*aṣṭāṅga-yoga*)[34] dealing with the physical, moral, psychological, and spiritual dimensions of the yogin, an integral path that emphasizes organic continuity, balance, and integration in contrast to the discontinuity, imbalance, and disintegration inherent in *saṃyoga*. The idea of cosmic balance and of the mutual support and upholding of the various parts of nature and society is not foreign to Yoga thought. Vyāsa deals with the theory of "nine causes" (*nava kāraṇāni*) or types of causation according to tradition.[35] The ninth type of cause is termed *dhṛti*—meaning "support" or "sustenance." Based on Vyāsa's explanation of *dhṛti* we can see how mutuality and sustenance are understood as essential conditions for the maintenance of the natural and social world. There is an organic interdependence of all living entities wherein all (i.e., the elements, animals, humans, and divine bodies) work together for the "good" of the whole and for each other.

Far from being exclusively a subjectively oriented and introverted path of withdrawal from life, classical Yoga acknowledges the intrinsic value of "support" and "sustenance" and the interdependence of all living (embodied) entities, thus upholding organic continuity, balance, and integration within the natural and social world. Having achieved that level of insight (*prajñā*) that is "truth-bearing" (*ṛtaṃbharā*),[36] the yogin perceives the natural order (*ṛta*) of cosmic existence, "unites" with, and embodies that order. To fail to see clearly (*adarśana*) is to fall into disorder, disharmony, and conflict with oneself and the world. In effect, to be ensconced in ignorance implies a disunion with the natural order of life and inextricably results in a failure to embody that order. Through Yoga one gains proper access to the world and is therefore established in right relationship to the world. Far from being denied or renounced, the world, for the yogin, has become transformed, properly engaged.

We need not read Patañjali as saying that the culmination of all yogic endeavor—*kaivalya*—is a static finality or inactive, isolated, solipsistic state of being. *Kaivalya* can be seen to incorporate an integrated, psychological consciousness along with the autonomy of pure consciousness, yet pure consciousness to which the realm of the *guṇas* (e.g., psychophysical being) is completely attuned and integrated. On the level of individuality, the yogin has found his (her) place in the world at large, "fitting into the whole."[37]

In the last chapter of the YS (*Kaivalya-Pāda*), "aloneness" (*kaivalya*) is said to ensue upon the attainment of *dharmamegha-samādhi*, the "cloud of dharma" *samādhi*. At this level of practice, the yogin has abandoned any search for (or attachment to) reward or "profit" from his or her meditational practice; a non-acquisitive attitude (*akusīda*) must take place at the highest level of yogic discipline.[38] Vyāsa emphasizes that the identity of *puruṣa* is not something to be acquired (*upādeya*) or discarded (*heya*).[39] The perspective referred to as "Pātañjala Yoga Darśana" culminates in a permanent state of clear "seeing" brought about through the discipline of Yoga. Yoga thus incorporates both an end state or "goal" and a process.[40]

Dharmamegha-samādhi presupposes that the yogin has cultivated higher dispassion (*para-vairāgya*)—the means to the enstatic consciousness realized in *asaṃprajñāta-samādhi*.[41] Thus, *dharmamegha-samādhi* is more or less a synonym of *asaṃprajñāta-samādhi* and can even be understood as the consummate phase of the awakening disclosed in enstasy, the final step on the long and arduous yogic journey to authentic identity and "aloneness."[42] A permanent identity shift—from the perspective of the human personality to *puruṣa*—takes place. Now free from any dependence on or subordination to knowledge or *vṛtti*, and detached from the world of misidentification (*saṃyoga*), the yogin yet retains the purified guṇic powers of virtue including illuminating "knowledge of all"[43] (due to purified *sāttva*), nonafflicted activity[44] (due to purified *rājas*), and a stable body-form (due to purified *tāmas*).

YS IV.30 declares: "From that [*dharmamegha-samādhi*] there is the cessation of afflicted action."[45] Hence the binding influence of the *guṇas* in the form of the afflictions, past actions, and misguided relationships is overcome; what remains is a "cloud of dharma" which includes an "eternality of knowledge" free from all impure covering (*āvaraṇa-mala*, YS IV.31) or veiling affliction and where "little (remains) to be known."[46] The eternality or endlessness of knowledge is better understood metaphorically rather than literally: It is not knowledge expanded to infinity but implies *puruṣa*-realization which transcends the limitations and particulars of knowledge (*vṛtti*).

The culmination of the Yoga system is found when, following from *dharmamegha-samādhi*, the mind and actions are freed from misidentification and affliction and one is no longer deluded/confused with regard to one's true form (*svarūpa*) or intrinsic identity. At this stage of practice the yogin is disconnected (*viyoga*) from all patterns of action motivated by the ego. According to both Vyāsa[47] and the sixteenth century commentator Vijñāna Bhikṣu,[48] one to whom this high state of purification takes place is designated as a *jīvanmukta*: one who is liberated while still alive (i.e., embodied or living liberation).

By transcending the normative conventions and obligations of karmic behavior, the yogin acts morally not as an extrinsic response and out of obedience to an external moral code of conduct, but as an intrinsic response and as a matter of natural, purified inclination. The stainless luminosity of pure consciousness is revealed as one's fundamental nature. The yogin does not act saṃsārically and ceases to act from the perspective of a delusive sense of self confined within *prakṛti's* domain. Relinquishing all obsessive or selfish concern with the results of activity, the yogin remains wholly detached from the egoic fruits of action.[49] This does not imply that the yogin loses all orientation for action. Only attachment (and compulsive, inordinate desire), not action itself, sets in motion the law of moral causation (karma) by which a person is implicated in *saṃsāra*. The yogin is said to be nonattached to either virtue or non-virtue, and is no longer oriented within the egological patterns of thought as in the epistemically distorted condition of *saṃyoga*. This does not mean, as some scholars have misleadingly concluded, that the spiritual adept or yogin is free to commit immoral acts,[50] or that the yogin is motivated by selfish concerns.[51]

Actions must not only be executed in the spirit of unselfishness (i.e., sacrifice) or detachment, they must also be ethically sound, reasonable and justifiable. Moreover, the yogin's spiritual journey—far from being an "a-moral process"[52]—is a highly moral process! The yogin's commitment to the sāttvification of consciousness, including the cultivation of moral virtues such as compassion (*karuṇā*)[53] and nonviolence (*ahiṃsā*),[54] is not an "a-moral" enterprise, nor is it an expression of indifference, aloofness, or an uncaring attitude to others. Moral disciplines are engaged as a natural outgrowth of intelligent (sāttvic) self-understanding, insight, and commitment to self-transcendence that takes consciousness out of (ec-stasis) its identification with the rigid structure of the monadic ego, thereby reversing the inveterate tendency of this ego to inflate itself at the expense of its responsibility in relation to others.

Having defined the "goal" of Yoga as "aloneness" (*kaivalya*), the question must now be asked: What kind of "aloneness" was Patañjali talking about? "Aloneness," I suggest, is not the isolation of the seer (*draṣṭṛ*, *puruṣa*) separate from the seeable (*dṛśya*, *prakṛti*), as is unfortunately far too often maintained as the goal of Yoga, but refers to the "aloneness" of the power of "seeing" (YS II.20, 25) in its innate purity and clarity without any epistemological distortion and moral defilement. The cultivation of *nirodha* uproots the compulsive tendency to reify the world and oneself (i.e., that pervading sense of separate ego irrevocably divided from the encompassing world) with an awareness that reveals the transcendent, yet immanent seer (*puruṣa*). Through clear "seeing" (*dṛśi*) the purpose of Yoga is fulfilled, and the yogin, free from all misidentification and impure karmic residue (as in the former contextual sphere of *cittavṛtti*), gains full, immediate access to the world. By accessing the world in such an open and direct manner, in effect "uniting" (epistemologically) with the world, the yogin ceases to be encumbered by egoism (i.e., *asmitā* and its egoic attitudes and identity patterns), which, enmeshed in conflict and confusion and holding itself as separate from the world, misappropriates the world.

Yoga can be seen to unfold—in *samādhi*—states of epistemic oneness that reveal the non-separation of knower, knowing, and the known (YS I.41) grounding our identity in a nonafflicted mode of action. *Kaivalya* implies a power of "seeing" in which the dualisms rooted in our egocentric patterns of attachment, aversion, fear, and so forth have been transformed into unselfish ways of being with others.[55] The psychological, ethical, and social implications of this kind of identity transformation are, needless to say, immense. I am suggesting that Yoga does not destroy or anesthetize our feelings and emotions thereby encouraging neglect and indifference toward others. On the contrary, the process of "cessation" (*nirodha*) steadies one for a life of compassion, discernment, and service informed by a "seeing" that is able to understand (literally meaning "to stand among, hence observe")—and is in touch with—the needs of others. What seems especially relevant for our understanding of Yoga ethics is the enhanced capacity generated in Yoga for empathic identification with the object one seeks to understand. This is a far cry from the portrayal of the yogin as a disengaged figure, psychologically and physically removed from the human relational sphere, who in an obstinate and obtrusive fashion severs all ties with the world. Such an image of a wise yogin merely serves to circumscribe our vision of humanity and, if anything else, stifle the spirit by prejudicing a spiritual, abstract (and disembodied) realm over and against nature and our human embodiment. In Yoga philosophy "seeing" is not only a cognitive term but implies purity of mind, that is, it has moral content and value. Nor is "knowledge" (*jñāna, vidyā*) in the Yoga tradition to be misconstrued as a "bloodless" or "heartless" gnosis.

This paper therefore suggests that through the necessary transformation of consciousness brought about in *samādhi*, an authentic and fruitful coherence of self-identity, perception, and activity emerges out of the former fragmented consciousness in *saṃyoga*. If Patañjali's perception of the world of forms and differences had been destroyed or discarded, how could he have had such insight into Yoga and the intricacies and subtle nuances of the unenlightened state?[56] If through *nirodha* the individual form and the whole world had been canceled for Patañjali, he would more likely have spent the rest of his days in the inactivity and isolation of transcendent oblivion rather than present Yoga philosophy to others! Rather than being handicapped by the exclusion of thinking, perceiving, experiencing, or activity, the liberated yogin actualizes the potential to live a fully integrated life in the world. I conclude here that there is no reason why the liberated yogin cannot be portrayed as a vital, creative, thoughtful, empathetic, balanced, happy, and wise person. Having adopted an integrative orientation to life, the enlightened being can endeavor to transform, enrich, and ennoble the world. I am therefore suggesting that there is a rich affective, moral, and cognitive as well as spiritual potential inherent in the realization of *puruṣa*, the "aloneness" of the power of consciousness/seeing.

Yoga presupposes the integration of knowledge and activity; there can be no scission between theoria and praxis. The *Yoga Sūtra* is a philosophical text where praxis is deemed to be essential. Without actual practice the theory that informs Yoga would have no authentic meaning. Yet without examination and reflection there would be no meaningful striving for liberation, no "goal," as it were, to set one's sight on. In an original, inspiring, and penetrating style, Patañjali bridges metaphysics and ethics, transcendence and immanence, and contributes to the Hindu fold a form of philosophical investigation that, to borrow J. Taber's descriptive phrase for another context, can properly be called a "transformative philosophy." That is to say, it is a philosophical perspective which "does not stand as an edifice isolated from experience; it exists only insofar as it is realized in experience."[57]

Conclusion

To conclude, it can be said that *puruṣa* indeed has some precedence over *prakṛti* in Patañjali's system, for *puruṣa* is what is ordinarily "missing" or concealed in human life and is ultimately the state of consciousness one must awaken to in Yoga. The liberated state of "aloneness" (*kaivalya*) need not denote either an ontological superiority of *puruṣa* or an exclusion of *prakṛti. Kaivalya* can be positively construed as an integration of both principles—an integration that, I have argued, is what is most important for Yoga. I have proposed that the *Yoga Sūtra* does not uphold a "path" of liberation that ultimately renders *puruṣa* and *prakṛti* incapable of "co-operating" together. Rather, the *Yoga Sūtra* seeks to "unite" these two principles without the presence of any defiled understanding, to bring them "together," properly aligning them in a state of balance, harmony, and a clarity of knowledge in the integrity of being and action.

The purified mind, one that has been transformed through yogic discipline, is certainly no ordinary worldly awareness nor is it eliminated for the sake of pure consciousness. To confuse (as many interpretations of Yoga have unfortunately done) the underlining purificatory processes involved in the cessation of ignorance/afflicted identity as being the same thing as (or as necessitating the need for) a radical elimination of our psychophysical being—the prakṛtic vehicle through which consciousness discloses itself—is, I suggest, to misunderstand the intent of the *Yoga Sūtra* itself. There are strong grounds for arguing (as I have done) that through "cessation" *prakṛti* herself (in the form of the *guṇic* constitutional makeup of the yogin's body-mind) is liberated from the grip of ignorance. Vyāsa explicitly states (*YB* II.18) that emancipation happens in the mind and does not literally apply to *puruṣa*—which is by definition already free and therefore has no intrinsic need to be released from the fetters of saṃsāric existence.

Both morality and perception (cognition) are essential channels through which human consciousness, far from being negated or suppressed, is transformed and illuminated. Yoga combines discerning knowledge with an emotional, affective, and moral sensibility allowing for a participatory epistemology that incorporates the moral amplitude for empathic identification with the world, that is, with the objects or persons one seeks to understand. The enhanced perception gained through Yoga must be interwoven with Yoga's rich affective and moral dimensions to form a spirituality that does not become entangled in a web of antinomianism, but which retains the integrity and vitality to transform our lives and the lives of others in an effective manner. In Yoga proper there can be no support, ethically or pedagogically, for the misappropriation or abuse of *prakṛti* for the sake of freedom or *puruṣa*-realization. By upholding an integration of the moral and the mystical, Yoga supports a reconciliation of the prevalent tension within Hinduism between (1) spiritual engagement and self-identity within the world (*pravṛtti*) and (2) spiritual disengagement from worldliness and self-identity that transcends the world (*nivṛtti*). Yoga discerns and teaches a balance between these two apparently conflicting orientations.

This paper has attempted to counter the radically dualistic, isolationistic, and ontologically oriented interpretations of Yoga presented by many scholars—where the full potentialities of our human embodiment are constrained within a radical, rigid, dualistic metaphysical structure—and propose instead an open-ended, morally and epistemologically oriented hermeneutic that frees Yoga of the long-standing conception of spiritual isolation, disembodiment, self-denial, and world-negation and thus from its pessimistic image. Our interpretation does not impute that *kaivalya* denotes a final incommensurability between spirit and matter. While Patañjali can be understood as having adopted a provisional, practical, dualistic metaphysics, there is no proof that his system either ends in duality or eliminates the possibility for an ongoing cooperative duality. Yoga is not simply "*puruṣa*-realization"; it equally implies "getting it right with *prakṛti*".

As well as being one of the seminal texts on yogic technique and transformative/liberative approaches within Asian Indian philosophy, Patañjali's *Yoga Sūtra* has to this day remained one of the most influential spiritual guides in Hinduism. In addition to a large number of people within India, millions of Westerners are actively practicing some form of Yoga influenced by Patañjali's thought clearly demonstrating Yoga's relevance for today as a discipline that can transcend cultural, religious, and philosophical barriers. The universal and universalizing potential of Yoga makes it one of India's finest contributions to our struggle for self-definition, moral integrity, and spiritual renewal today. The main purpose of this paper has been to consider a fresh approach in which to reexamine and reassess classical Yoga philosophy, and to help to articulate in a fuller way what I have elsewhere referred to as the integrity of the Yoga Darśana.[58] Thus, it is my hope that some of the suggestions presented here can function as a catalyst for bringing Patañjali's thought into a more fruitful dialogue and encounter with other religious and philosophical traditions both within and outside of India.

Endnotes

[1] The system of classical Yoga is often reduced to or fitted into a classical Sāṃkhyan scheme—the interpretations of which generally follow along radically dualistic lines. In their metaphysical ideas classical Sāṃkhya and Yoga are closely akin. However, both systems hold divergent views on important areas of doctrinal structure such as epistemology, ontology, ethics, and psychology, as well as differences pertaining to terminology. These differences derive in part from the different methodologies adopted by the two schools: Sāṃkhya, it has been argued, emphasizes a theoretical or intellectual analysis through inference and reasoning in order to bring out the nature of final emancipation, while Yoga stresses yogic perception and multiple forms of practice that culminate in *samādhi*. Moreover, there is clear evidence throughout all four *pādas* of the YS of an extensive network of terminology that parallels Buddhist teachings and which is absent in the classical Sāṃkhya literature. Patañjali includes several sūtras on the "restraints" or *yamas* (namely, nonviolence [*ahiṃsā*], truthfulness [*satya*], non-stealing [*asteya*], chastity [*brahmacarya*], and nonpossession [*aparigraha*]) of the "eight-limbed" path of Yoga that are listed in the *Acārāṅga Sūtra* of Jainism (the earliest sections of which may date from the third or fourth century B.C.E.) thereby suggesting possible Jaina influences on the Yoga tradition. The topic of Buddhist or Jaina influence on Yoga doctrine (or vice versa) is, however, not the focus of this paper.

[2] See, for example, Śaṅkara's (ca eighth-ninth century CE) use of *vyāvahārika* (the conventional empirical perspective) in contrast to *paramārthika* (the ultimate or absolute standpoint).

[3] See Whicher (1998).

[4] See in particular: Feuerstein (1980: 14, 56, 108); Eliade (1969: 94–95, 99–100); Koelman (1970: 224, 251); and G. Larson (1987: 13) who classifies Patañjali's Yoga as a form of Sāṃkhya.

---

[5] F. Edgerton (1924), "The Meaning of Sāṃkhya and Yoga," AJP 45, pp. 1–46.

[6] As argued in Whicher (1998).

[7] YS I.2 (p. 4): *yogaś cittavṛttinirodhaḥ.* The Sanskrit text of the YS of Patañjali and the YB of Vyāsa is from The Yoga-Sūtras of Patañjali (1904), K. S. Āgāśe ed. (Poona: Ānandāśrama) Sanskrit Ser. no. 47. The modifications or functions (*vṛtti*) of the mind (*citta*) are said to be fivefold (YS I.6), namely, 'valid cognition' (*pramāṇa*, which includes perception [*pratyakṣa*], inference [*anumāna*] and valid testimony [*āgama*]), 'error'/'misconception' (*viparyaya*), 'conceptualization' (*vikalpa*), 'sleep' (*nidrā*) and 'memory' (*smṛti*), and are described as being 'afflicted' (*kliṣṭa*) or 'nonafflicted' (*akliṣṭa*) (YS I.5). *Citta* is an umbrella term that incorporates 'intellect' (*buddhi*), 'sense of self' (*ahaṃkāra*) and 'mind-organ' (*manas*), and can be viewed as the aggregate of the cognitive, conative and affective processes and functions of phenomenal consciousness, i.e., it consists of a grasping, intentional and volitional consciousness. For an in-depth look at the meaning of the terms *citta* and *vṛtti* see I. Whicher (1997, 1998). "The Mind (*Citta*): Its Nature, Structure and Functioning in Classical Yoga." in *Saṃbhāṣā* Vols 18 (pp. 35–62) and 19 (1–50). In the first four sūtras of the first chapter (*Samādhi-Pāda*) the subject matter of the YS is mentioned, defined and characterized. The sūtras run as follows: YS I.1: "Now [begins] the discipline of Yoga." YS I.2: "Yoga is the cessation of [the misidentification with] the modifications of the mind." YS I.3: "Then [when that cessation has taken place] there is abiding in the seer's own form (i.e., *puruṣa* or intrinsic identity)." YS I.4: "Otherwise [there is] conformity to (i.e., misidentification with) the modifications [of the mind]." YS I.1–4 (pp. 1, 4, 7, and 7 respectively): *atha yogānuśāsanam; yogaś cittavṛttinirodhaḥ; tadā draṣṇuḥ svarūpe'vasthānam; vṛttisārūpyam itaratra.* For a more comprehensive study of classical Yoga including issues dealt with in this paper see Whicher (1998) *The Integrity of the Yoga Darśana* (SUNY Press).

[8] See Whicher (1997, 1998).
[9] See Whicher (1997, 1998).
[10] See chapter 6 in Whicher (1998).
[11] YS II.15 (p. 74): *pariṇāmatāpasaṃskāraduḥkhair guṇavṛttivirodhāc ca duḥkham eva sarvaṃ vivekinaḥ*." Because of the dissatisfaction and sufferings due to change and anxieties and the latent impressions, and from the conflict of the modifications of the *guṇas*, for the discerning one, all is sorrow alone."
[12] Patañjali uses the term *pratiprasava* twice, in YS II.10 and IV.34.
[13] See Chapple and Kelly (1990) p. 60.
[14] Feuerstein (1979a) p. 65.
[15] Cf. T. Leggett (1990) p. 195 and U. Arya (1986) pp. 146, 471.
[16] The term *kaivalya* comes from *kevala*, meaning 'alone'. Feuerstein (1979a: 75) also translates *kaivalya* as "aloneness" but with a metaphysical or ontological emphasis that implies the absolute separation of *puruṣa* and *prakṛti*.
[17] YS II.25 (p. 96): *tadabhāvāt saṃyogābhāvo hānaṃ taddṛśeḥ kaivalyaṃ*.
[18] YS II.20 and IV.18.
[19] YS IV.34 (p. 207): *puruṣārthaśūnyānāṃ guṇānāṃ pratiprasavaḥ kaivalyaṃ svarūpapratiṣñhā vā citiśaktir iti*.
[20] See n. 19 above.
[21] YS III.55 (p. 174): *sattvapuruṣayoḥ śuddhisāmye kaivalyam iti*. One must be careful not to characterize the state of *sattva* itself as liberation or *kaivalya*, for without the presence of *puruṣa* the mind (as reflected consciousness) could not function in its most transparent aspect as sattva. It is not accurate, according to Yoga philosophy, to say that the *sattva* is equivalent to liberation itself. The question of the nature of the guṇas from the enlightened perspective is an interesting one. In the *Bhagavadgītā* (II.45) Kṛṣṇa advises Arjuna to become free from the three *guṇas* and then gives further instructions to be established in eternal *sattva* (beingness, light, goodness, clarity,

knowledge), free of dualities, free of acquisition-and-possession, Self-possessed (*nirdvandvo nityasattvastho niryogakṣema ātmavān*). It would appear from the above instructions that the nature of the *sattva* being referred to here transcends the limitations of the nature of *sattva-guṇa* which can still have a binding effect in the form of attachment to joy and knowledge. It is, however, only by first overcoming rajas and *tamas* that liberation is possible.

[22] YB III.55 (p. 175): nahi *dagdhakleśabījasya jñāne punar apekṣā kācid asti*. "When the seeds of afflictions have been scorched there is no longer any dependence at all on further knowledge."

[23] H. Āraṇya writes (1963: 123) that in the state of *nirodha* the *guṇas* "do not die out but their unbalanced activity due to non-equilibrium that was taking place ... only ceases on account of the cessation of the cause (*avidyā* or nescience) which brought about their contact."

[24] YB IV.25 (p. 201): *puruṣas tv asatyām avidyāyāṃ śuddhaś cittadharmair aparāmṛṣña*.

[25] YB I.41.

[26] YS II.26.

[27] YS III.49.

[28] Vijñāna Bhikṣu insists (YV IV.34: 141) that *kaivalya* is a state of liberation for both *puruṣa* and *prakṛti* each reaching its respective natural or intrinsic state. He then cites the Sāṃkhya-Kārikā (62) where it is stated that no *puruṣa* is bound, liberated or transmigrates. It is only *prakṛti* abiding in her various forms that transmigrates, is bound and becomes liberated. For references to Vijñāna Bhikṣu's YV I have consulted T. S. Rukmani (1981, 1983, 1987, 1989).

[29] YS I.51 and III.8; the state of *nirbīja* or "seedless" *samādhi* can be understood as the liberated state where no "seed" of ignorance remains, any further potential for affliction (i.e., as mental impressions or *saṃskāras*) having been purified from the mind.

[30] RM I.1 (p. 1).

[31] Müller (1899: 309).

[32] See, for example, Eliade (1969), Koelman (1970), Feuerstein (1979a), and Larson (1987).

[33] I am here echoing some of the points made by Chapple in his paper entitled,"*Citta-vṛtti* and Reality in the *Yoga Sūtra*" in *Sāṃkhya-Yoga: Proceedings of the IASWR Conference*, 1981 (Stoney Brook, New York: The Institute for Advanced Studies of World Religions, 1983), pp. 103–119. See also Chapple and Kelly (1990: 5) where the authors state: " ... *kaivalyam* ... is not a catatonic state nor does it require death." SK 67 acknowledges that even the "potter's wheel" continues to turn because of the force of past impressions (*saṃskāras*); but in Yoga, higher dispassion and *asaṃprajñāta* eventually exhaust all the impressions or karmic residue. Through a continued program of ongoing purification Yoga allows for the possibility of an embodied state of freedom utterly unburdened by the effects of past actions. As such Yoga constitutes an advance over the fatalistic perspective in Sāṃkhya where the "wheel of *saṃsāra*" continues (after the initial experience of liberating knowledge) until, in the event of separation from the body, *prakṛti* ceases and unending "isolation" (*kaivalya*) is attained (SK 68). In any case, the yogic state of supracognitive *samādhi* or enstasy goes beyond the liberating knowledge of *viveka* in the Sāṃkhyan system in that the yogin must develop dispassion even toward discriminative discernment itself. For more on an analysis of the notion of liberation in Sāṃkhya and Yoga see C. Chapple's chapter on "Living Liberation in Sāṃkhya and Yoga" in *Living Liberation in Hindu Thought*, ed. by Andrew O. Fort and Patricia Y. Mumme (Albany: State University of New York Press, 1996).

[34] YS II.29; see the discussion on *aṣṭāṅga-yoga* in chapter 4 of Whicher (1998).

[35] YB II.28 (pp. 99–101).

[36] YS I.48.

[37] See K. Klostermaier (1989), "Spirituality and Nature" in *Hindu Spirituality: Vedas Through Vedānta* ed. by Krishna Sivaraman (London: SCM Press) pp. 319–337.

[38] YS IV.29 (p. 202): *prasaṃkhyāne'py akusīdasya sarvathā vivekakhyāter dharmameghaḥ samādhiḥ.*

[39] YB II.15 (p. 78): *tatra hātuḥ svarūpamupādeyaṃ vā heyaṃ vā na bhavitumarhati.* "Here, the true nature/identity of the one who is liberated cannot be something to be acquired or discarded."

[40] Thus the term "Yoga" (like the terms "*nirodha*" and "*samādhi*") is ambiguous in that it means both the process of purification and illumination and the final result of liberation or "aloneness." Due to Yoga's traditional praxis-orientation it becomes all too easy to reduce Yoga to a "means only" approach to well-being and spiritual enlightenment. In the light of its popularity in the Western world today in which technique and practice have been emphasized often to the exclusion of philosophical/theoretical understanding and a proper pedagogical context, there is a great danger in simply reifying practice whereby practice becomes something the ego does for the sake of its own security. Seen here, practice—often then conceived as a superior activity in relation to all other activities—becomes all-important in that through the activity called "practice" the ego hopes and strives to become "enlightened." Practice thus becomes rooted in a future-oriented perspective largely motivated out of a fear of not becoming enlightened; it degenerates into a form of selfishly appropriated activity where "means" become ends-in-themselves. Moreover, human relationships become instruments for the greater "good" of Self-realization. Thus rationalized, relationships are seen as having only a tentative nature. The search for enlightenment under the sway of this kind of instrumental rationality/reasoning (that is, the attempt to "gain" something from one's practice, i.e., enlightenment) never really goes beyond the level of ego and its compulsive search for permanent security which of

course, according to Yoga thought, is an inherently afflicted state of affairs. To be sure, the concern of Yoga is to (re)discover *puruṣa*, to be restored to true identity thus overcoming dissatisfaction, fear and misidentification by uprooting and eradicating the dis-ease of ignorance (*avidyā*). Yet, as W. Halbfass puts it, true identity "cannot be really lost, forgotten or newly acquired" (1991: 252) for liberation "is not to be produced or accomplished in a literal sense, but only in a figurative sense" (ibid: 251). Sufficient means for the sattvification of the mind are, however, both desirable and necessary in order to prepare the yogin for the necessary identity shift from egoity to *puruṣa*. By acknowledging that "aloneness" cannot be an acquired state resulting from or caused by yogic methods and techniques, and that *puruṣa* cannot be known (YB III.35), acquired or discarded/lost (YB II.15), Yoga in effect transcends its own result-orientation as well as the categories of means and ends.

[41] YB I.18.

[42] See Feuerstein (1980: 98).

[43] YS III.49 and III.54.

[44] YS IV.7; see also YS IV.30 (n. 45 below).

[45] YS IV.30 (p. 202): *tataḥ kleśakarmanivṛttiḥ*. Thus, it may be said that to dwell without defilement in a "cloud of dharma" is the culminating description by Patañjali of what tradition later referred to as living liberation (*jīvanmukti*). To be sure, there is a "brevity of description" in the YS regarding the state of liberation. Only sparingly, with reservation (one might add, caution) and mostly in metaphorical terms does Patañjali speak about the qualities exhibited by the liberated yogin. Chapple (1996: 116, see below) provides three possible reasons for this "brevity of description" regarding living liberation in the context of the YS (and Sāṃkhya, i.e. the SK of Īśvara Kṛṣṇa): (1) He states: "(T)he genre in which both texts were written does not allow for the sort of narrative and poetic embellishment found in the epics and Purāṇas." (2) Perhaps, as Chapple suggests "...

a deliberate attempt has been made to guarantee that the recognition of a liberated being remains in the hands of a spiritual preceptor." What is to be noted here is that the oral and highly personalized lineage tradition within Yoga stresses the authority of the guru which guards against false claims to spiritual attainment on the part of others and thereby "helps to ensure the authenticity and integrity of the tradition." (3) A further reason for brevity "could hinge on the logical contradiction that arises due to the fact that the notion of self is so closely identified with *ahaṃkāra* [the mistaken ego sense or afflicted identity]. It would be an oxymoron for a person to say [']I am liberated.[']" The Self (*puruṣa*) is of course not an object which can be seen by itself thus laying emphasis, as Chapple points out, on the ineffable nature of the liberative state which transcends mind-content, all marks and activity itself.

[46] YS IV.31 (p. 203): *tadā sarvāvaraṇamalāpetasya jñānasyānantyājjñeyam alpam.*

[47] See YB IV.30 (pp. 202–203): *kleśakarmanivṛttau jīvanneva vidvānvimukto bhavati.* On cessation of afflicted action, the knower is released while yet living."

[48] YV IV.30 (pp. 123–124). Elsewhere in his *Yoga-Sāra-Saṃgraha* (p. 17) Vijñāna Bhikṣu tells us that the yogin who is "established in the state of *dharmamegha-samādhi* is called a *jīvanmukta*" (... *dharmameghaḥ samādhiḥ* ... *asyāmavasthāyāṃ jīvanmukta ityucyate*). Vijñāna Bhikṣu is critical of Vedāntins (i.e. Śaṅkara's Advaita Vedānta school) that, he says, associate the *jīvanmukta* with ignorance (*'avidyā-kleśa'*)—probably because of the liberated being's continued link with the body—despite Yoga's insistence on the complete overcoming of the afflictions.

[49] This is the essence of Kṛṣṇa's teaching in the Bhagavadgītā on *karmayoga*; see, for example, BG IV.20.

[50] See R. C. Zaehner (1974), *Our Savage God* (London: Collins) pp. 97–98.

---

[51] See B.-A. Scharfstein (1974), *Mystical Experience* (Baltimore, MD: Penguin) pp. 131–132.
[52] See Feuerstein (1979a: 81).
[53] YS I.33 (p. 38): *maitrīkaruṇāmuditopekṣāṇāṃ sukhaduḥkhapuṇyāpuṇyaviṣayāṇāṃ bhāvanātaś cittaprasādanam.* "The mind is made pure and clear from the cultivation of friendliness, compassion, happiness and equanimity in conditions or toward objects of joy, sorrow, merit or demerit respectively."
[54] YS II.35.
[55] YS I.33; see n. 53 above.
[56] Although the historical identity of Patañjali the Yoga master is not known, we are assuming that Patañjali was, as the tradition would have it, an enlightened Yoga adept.
[57] J. Taber (1983). *Transformative Philosophy: A Study of Śaṅkara, Fichte and Heidegger* (Honolulu: University of Hawaii Press) p. 26.
[58] See Whicher (1998).

Super *Saṃskāras*: Soteriological Subliminal Impressions in Patañjali's *Yoga Sūtra*

Beverley Foulks[1]

Early Indian texts, including the *Rig Veda* (10.136) and the *Upaniṣads* (*Kaṭha Upaniṣad* II 3.11), mention Yoga as an ascetic discipline aimed at controlling one's senses, but Patañjali's *Yoga Sūtra* is the first text within the Hindu tradition that is entirely devoted to describing the practice and philosophy of Yoga. Patañjali's *Yoga Sūtra* was written in approximately the third century CE, followed by a commentary entitled the *Yoga-Bhāṣya* written by Vyāsa in the eighth century CE. The word Yoga comes from the Sanskrit root word *yuj*, meaning to "unite, join, or connect." Within Patañjali's text, Yoga is described first and foremost as "the cessation of the turnings of thought (*citta-vṛtti*)" (I.2).[2] As an active performance (*kriyā-yoga*), it involves "ascetic practice (*tapas*), study of sacred lore (*svādhyāya*), and dedication to the Lord of Yoga (*īśvara-praṇidhāna*)" (II.1). Its purpose is "to cultivate pure contemplation and attenuate the forces of corruption" (II.23). The eight limbs of Yoga includes moral principles (*yama*), observances (*niyama*), posture (*āsana*), breath control (*prāṇāyāma*), withdrawal of the senses (*pratyāhāra*), concentration (*dhāraṇa*), meditation (*dhyāna*), and pure contemplation (*samādhi*)" (II.29). In this paper, I focus primarily on the first definition of Yoga as "cessation of the turnings of thought (*citta-vṛtti*)." Specifically, I argue that Patañjali affords a new role to be played by *saṃskāra*s (subliminal impressions) in bringing about cessation of such turnings of thought. Not only do *saṃskāra*s perform their traditional function of perpetuating such turnings of thought along with *vāsana*s (memory traces) and *āśaya*s (subliminal intentions), but one particular type of *saṃskāra* generated by wisdom (*prajñā*) has the ability to bring about a cessation of turnings of thought altogether. This *saṃskāra* has supreme soteriological significance within Patañjali's system of Yoga, as it represents the mechanism through which the *puruṣa* might cease identifying with turnings of thought and become able to observe the world unobstructed.

In recent studies of Patañjali's *Yoga Sūtra,* scholars have been interested primarily in the issue of whether or not cessation (*nirodha*) entails the *puruṣa* isolating itself entirely from *prakṛti,* or whether the *puruṣa* somehow remains engaged in the world in spite of cessation of turnings of thought. Whereas some scholars have argued that Patañjali's system follows the dualism of *puruṣa* and *prakṛti* as suggested by Sāṃkhya philosophy[3], or even that his Yoga is the practical implementation of Sāṃkhya's intellectual project[4], others insist that one misinterprets what is meant by *kaivalya* (translated as "aloneness" by Ian Whicher, "freedom" by Barbara Stoler Miller) if one insists that *puruṣa* is completely isolated from *prakṛti* following cessation.[5] Much is at stake in these debates. The former position insists that a *yogin/yoginī* who has attained cessation under Patañjali's system of Yoga actually transcends or dissolves the material world, while the latter position maintains that the *yogin/yoginī* can play an active role in that world even after such attainments. Those who would like to maintain the latter position must therefore explain how one who has attained "seedless contemplation" (*nīrbija-samādhi*) or the "essential cloud" (*dharma-megha*) engages with *prakṛti,* since the *yogin/yoginī* has at that point superseded ordinary consciousness wherein experiences generate subliminal impressions (*saṃskāra*s) that lie dormant until those memory traces (*vāsana*s) and subliminal intentions (*āśaya*s) then impel him/her to further act in the world.

In *The Integrity of the Yoga Darśaṇa*, Ian Whicher insists that what dissolves in *nirodha* is not the material world (*prakṛti*) but instead the misidentification of *puruṣa* with *prakṛti.*[6] He argues that the *puruṣa*, which was formerly eclipsed by mental functions and tangled in a net of impressions (*saṃskāra*s), habit patterns (*vāsāna*s) and turnings of thought (*citta-vṛtti*s), instead becomes revealed as the true seer after the arising of wisdom (*prajñā*).[7] Thus Whicher claims that cessation is not ontological, but instead epistemological.[8] While I would agree with Whicher's emphasis that the *yogin/yoginī* who has achieved cessation does not have a deadened mind or exist in an inactive state[9], I disagree with his claim that following *nirodha,* the *yogin/yoginī* still experiences turnings of thought (*citta-vṛtti*) but no longer misidentifies *puruṣa* with a transforming *prakṛti*. In my mind, this would contradict Patañjali's primary claim that Yoga involves a cessation of such turnings of thought. Thus I must challenge Whicher when, discussing the commentary by Vyāsa concerning *saṃskāra,* he states that:

> When Vyāsa tells us that the *saṃskāras* of the yogin have become "burned," their seeds "scorched" (*YB* IV.28), it is meant that *avidyā*—the cause of *saṃyoga* and all the *saṃskāras* that take the form of habit patterns of misconception or error (*viparyaya-vāsanās, YB* II.24)—has been burned, and not that the mind or "consciousness-of" (the power of identification) or the functioning of mental impressions and memory in total have been destroyed. The yogin's understanding and the functioning (*vṛtti)* of the mind have been transformed, not negated.[10]

This, I would claim, is a misinterpretation of the function and role of *saṃskāra*s in Patañjali's system. Patañjali says unequivocally that turnings of thought (*citta-vṛtti*)—which are fueled by *saṃskāra*s, *vāsanā*s, etc.—cease in Yoga. Without such cessation, the *yogin/yoginī* would not achieve seed-bearing contemplation (*sabīja-samādhi*) and seedless contemplation (*nirbīja-samādhi*). The *puruṣa* does not reveal itself as a seer of the world prior to such contemplations. Seedless contemplation (*nirbīja-samādhi*) means that there are no longer any seeds (*bīja*) generated from experience, thus turnings of thought (*citta-vṛtti*) cease completely.

By insisting upon the centrality of *saṃskāra*s within Patañjali's system, I hope to redress this prominent issue in scholarship of the *Yoga Sūtra*. Instead of interpreting the final state of seedless contemplation (*nirbīja-samādhi*) as a transcendent *puruṣa* appreciating the play of *prakṛti* while continuing to be subject to *citta-vṛtti*[11] or a non-dual experiencing of the world[12], I would emphasize that Patañjali does make a distinction between the "seeing" of the *puruṣa* and the experience of consciousness when entangled in turnings of thought. Yet I would similarly challenge scholars such as Lloyd Pflueger who insist that Patañjali's *sūtra* articulates a clear dualistic distinction between *puruṣa* and *prakṛti*.[13] I agree with Whicher when he claims that the deluded state of consciousness involves an epistemological distortion whereby the practitioner identifies mistakenly with *prakṛti*. In my interpretation of the *Yoga Sūtra,* it is *prakṛti* that the *puruṣa* observes following such cessation of turnings of thought. Yet the way that the *puruṣa* experiences the world differs markedly from the way that the *yogin/ yoginī* previously engaged in the world. The *puruṣa* does not dissolve or transcend the world entirely; instead, the revelation of the *puruṣa* as seer signals a transformation whereby the turnings of thought (*citta-vṛtti*) have ceased completely.

In order to better appreciate this transformation, I would suggest that we pay close attention to the way that Patañjali uses *saṃskāra*s within his project. Whicher argues that we should see *kaivalya* (aloneness, freedom) as a "bridge" concept that describes a *yogin* who is "*in* the world but is not defined by worldly existence."[13] He writes that *kaivalya* succeeds in "bridging together or harnessing two formerly undisclosed principles or powers by correcting a misalignment between them based on a misconception or misperception of authentic identity."[15] I would advocate instead that we adopt *saṃskāra* as a "bridge" concept, for as we will see, it is the "super *saṃskāra*" generated by wisdom that ultimately reveals the *puruṣa* as seer and brings about a cessation of all previous *saṃskāra* that have perpetuated turnings of thought (*citta-vṛtti*) and previously obscured the *puruṣa*. Whereas Whicher would like to reinterpret *kaivalya* to allow for the engagement of the *yogin/yoginī* in the world, I would argue that *saṃskāra* instead corrects the misalignment between the previously obscured *puruṣa* and the previously obscuring *prakṛti*.

Granted, this does not address the concern of Whicher and others who would like to maintain the possibility that a *yogin/yoginī* who has revealed his/her *puruṣa* can remain fully engaged in the world, but, as I will argue, it *does* prevent any interpretation of the *yogin/yoginī* as void of all mental activity or contact with *prakṛti* following cessation of the turnings of thought (*citta-vṛtti*). Unlike *kaivalya* (aloneness, freedom), which is a state that applies only following the cessation of such turnings of thought (*vṛtti*), *saṃskāra* is not only a concept that characterizes the state in which the *puruṣa* is obscured as the observer of the world because of *citta-vṛtti*, but it also functions in bringing about the cessation of *citta-vṛtti* and the revelation of the *puruṣa*. It, rather than *kaivalya*, better describes the achievement accomplished by Yoga in bringing about freedom from an afflicted or deluded consciousness.

In the *Yoga Sūtra*, subliminal impressions (*saṃskāras*) play a formative role in both the nature and practice of Yoga. On the one hand, subliminal impressions are all that remain following the conscious cessation of thought (I.18) in seed-bearing contemplation (*sabīja-samādhi*). Turnings of thought (*citta-vṛtti*), which occur in the mind, bind the spirit (*puruṣa*) to material nature (*prakṛti*) by leaving subliminal traces or impressions in the mind. Such subliminal impressions (*saṃskāras*) and memory traces (*vāsanās*)—the aftermath of all experiences—generate memories and further compel actions. Thus we can see that on the one hand, subliminal impressions (*saṃskāras*) and habit-patterns (*vāsanās*), rooted in the mind because of previous activities, bring about turnings of thought (*citta-vṛtti*). They are what David Carpenter calls the practitioner's "karmic stockpile" (*karmāśaya*).[16] The five types of afflictions that produce such *saṃskāra* are ignorance (*avidyā*), egoism (*asmitā*), desire (*raga*), hatred (*dveṣa*), and attachment to life (*abhiniveśa*). Such experiences fuel further actions, and the cycle of *samsāra* continues unabated. In his *Yoga Sūtra,* Patañjali details five turnings of thought (*vṛtti*): valid judgment (*pramāṇa*), error (*viparyaya*), conceptualization (*vikalpa*), sleep, and memory (*smṛti*) (I.6). Thus on the one hand, Patañjali describes *saṃskāra*s as those subliminal impressions that promulgate actions and turnings of thought, such that the *puruṣa* is obscured.

Yet Patañjali also describes a *saṃskāra,* generated by wisdom, which has the capacity to stop the formation of other subliminal impressions (II.50). It is this *saṃskāra,* which Vyāsa calls *prajñākṛtasaṃskāra*, which brings about contemplation without seeds (*nirbīja-samādhi*). Wisdom calms the mind completely, such that all mental activity and subliminal impressions cease, and the *yogin/yoginī* can perceive the world from the perspective of the *puruṣa.* Thus *saṃskāra*s not only play a negative role in Patañjali's *sūtra*, but this particular *saṃskāra* is essential for bringing about cessation so that the practitioner is spiritually liberated. David Carpenter suggests that such positive *saṃskāra*s are virtues within Patañjali's system.[17] Indeed, scholars have noted the seemingly paradoxical role of *saṃskāra* here; Lakshmi Kapani suggests that *saṃskāra*s have an ambivalent status in the *Yoga Sūtra* since they play a role in both bondage and liberation.[18] In attributing a dual role to *saṃskāra,* Patañjali breaks from traditional understandings of *saṃskāra*, both in Vedic literature in which it refers to ritual construction and purification and also in Buddhist literature where it only refers negatively to those subliminal impressions that further perpetuate karma and entrench a person in *samsāra.*

This notion of a *saṃskāra* generated by wisdom—what I am calling a "super *saṃskāra*"—is truly innovative within Patañjali's system. In the Hindu Dharma literature, *saṃskāra*s refer to rites that purify or perfect the human body (*śarīrasaṃskāras*).[19] These life-cycle rituals of birth, high-caste initiation (*upanayana*), marriage (*vivāha*), and funeral rites (*antyeṣṭi*) constitute those "occasional rituals" (*naimittika-karma*) that the Brahman householder must perform.[20] As Gavin Flood notes, *saṃskāra*s serve to construct social identities: "Indeed, the Sanskrit term for such rites is *saṃskāra*, 'constructed' or 'put together', implying the putting together of a person as a social actor and even, to some extent, defining ontological status. By undergoing the various *saṃskāras* a Hindu gains access to resources within the tradition which were previously closed to him or her and enters a new realm or state."[21]

In Buddhism, *saṃskāra* represents the fourth of the five aggregates (*skandhas*) of a human being, which are material form (*rūpa*), feeling or sensation (*vedanā*), perception or abstract thought (*saṃjñā*), mental formation/volition (*saṃskāra*), and consciousness (*vijñāna*). Yet as Har Dayal notes, the term "*saṃskāra*" is virtually untranslatable. Indeed, the term has been variously rendered as:

> Plastic forces, syntheses, pre-natal forces, potentialities, conformations, mental confections, conditions, precedent conditions, complexes of consciousness, activities, activities and capabilities, mental activities, actions, synergies, dispositions, predispositions, impressions, volitions, volitional complexes, constituents, mentations, aggregates, constituents of being, potencies, merit and demerit, mental co-efficients, mental qualities, actions, deeds, energies, productions…[22]

Given this expansive range of potential meanings for *saṃskāra*, it is striking that Patañjali should find yet another capacity for *saṃskāra* that supersedes this list. Although David

Carpenter suggests that the notion of *saṃskāra* as "disposition" is identical within Buddhism and Patañjali's Yoga system, the notion of a "super *saṃskāra*" has no precedent within Buddhism. As Louis de la Vallée Poussin remarks in his article, "Le Bouddhisme et le Yoga de Patañjali," there are over one hundred terms shared by Yoga and Buddhist philosophies, including *bīja, vāsanā,* and *āśaya*, but he notes that Patañjali's understanding of *saṃskāra* not only has no precedent in classic *darśana* texts, but it also has a different connotation from that of Buddhist texts.[23] Carpenter is correct in noting that *saṃskāra* has a decidedly psychological orientation within Patañjali's text. Within Buddhism, *saṃskāra*s are considered to be those subliminal impressions that determine one's future karma, including one's rebirth: thus they are most often called "mental formations" or "volitions." As this latter term suggests, *saṃskāra*s determine peoples' karmic inclinations. Yet Patañjali extends the capacity of *saṃskāra* beyond this internal mechanism within karma when he suggests that a "super *saṃskāra*" generated by wisdom might bring about complete cessation of turnings of thought.

When we examine what sorts of methods Patañjali describes as having the potential to bring about cessation of turnings of thought (*citta-vṛtti*), we find that *saṃskāra*s play an important role in these various means of achieving cessation. As Patañjali describes in the beginning of his *sūtra*, Yoga is the cessation of *citta-vṛtti* wherein the *puruṣa* stands in its true identity as observer of the world, unlike the previous tendency for the observer to identify with *citta-vṛtti* (I.2–4). There is both a negative and positive movement here: *citta-vṛtti* is stopped, and the *puruṣa* is revealed. The *Yoga Sūtra* focuses the majority of its attention on the first process rather than the second, perhaps because the revelation of the *puruṣa* is not only difficult to accomplish, but also difficult to conceptualize and articulate in words. By contrast, Patañjali describes a variety of ways to accomplish the cessation of *citta-vṛtti*. First, he suggests that one can bring about cessation through dispassion (*vairāgya*) and practice (*abhyāsa*). He describes the former as a mastery over craving of sensual objects and the latter as the attempt to maintain cessation. As we will later see, this becomes important: cessation of *citta-vṛtti* is only *fully* accomplished when such cessation can be maintained. Practice, specifically the seed-bearing (*sabīja*) and seedless (*nirbīja*) contemplations, is essential in order for the observer to cease identifying with *prakṛti* and the *puruṣa* ultimately to be revealed.

While Whicher argues that the *puruṣa* is still subject to *citta-vṛtti* after the attainment of cessation, I would insist that what is meant by "maintaining cessation" is not a repeated occurrence and cessation of *citta-vṛtti*, but instead the maintenance of cessation of those *citta-vṛttis* that further promulgate consciousness as well as the generation of "super *saṃskāra*" that stop the formation of other *saṃskāra*s. The point is subtle, but significant. Whereas Whicher suggests that the *puruṣa* operates with a mental faculty that is still subject to *saṃskāra* that are generated by *citta-vṛtti*, I would point to such descriptions of practice (*abhyāsa*) being the "maintenance of the cessation of thought" to argue that the *puruṣa* is only revealed once such *citta-vṛtti*, and the residual *saṃskāra*s generated from previous *citta-vṛtti*, cease altogether. Whereas dispassion enables the *yogin/yoginī* to begin stopping the *citta-vṛtti*, practice allows the practitioner to maintain that state.

When the function of practice (*abhyāsa*) and dispassion (*vairāgya*) is properly understood, we find that one can legitimately defend the notion of the *puruṣa* remaining active without suggesting that the *puruṣa* is still somehow subject to turnings of thought. In order for the *puruṣa* to be revealed, the practitioner must maintain the cessation through practice. Although the mental functioning of the mind has changed significantly, in that the *citta-vṛtti* have ceased, the practitioner must still exert him/herself in order to maintain this state of awareness. This becomes clear when we look at the discussion of Vyāsa about practice (*abhyāsa*) and dispassion (*vairāgya*) in his *Yoga-Bhāsya*. In his commentary (I.12) he writes:

> The stream of consciousness flows in both [directions]. It flows to the good, and it flows to the bad. The one commencing with discernment (*viveka*) and terminating in *kaivalya* flows to the good. The one commencing with lack-of-discernment (*aviveka*) and terminating in conditioned existence (*samsāra*) flows to the bad. Through dispassion (*vairāgya*) the flowing out to the sense objects is checked and through the practice (*abhyāsa*) of the vision of discernment the stream of discernment is laid bare. Thus the restriction of the fluctuations of consciousness is dependent upon both [*abhyāsa* and *vairāgya*].[24]

Here we see two movements at work: dispassion (*vairāgya*) terminates the flow of sense-objects and practice (*abhyāsa*) reveals the stream of discernment. Thus dispassion performs a negative act of stopping deluded thought, while practice enables the positive act of discernment. By suggesting that there are "two streams" at play, Vyāsa underscores the idea that Yoga involves not only a cessation of the bad, but also a cultivation of the good. What Vyāsa calls the "restriction of the fluctuations of consciousness" is that which we have called "cessation of the turnings of thought." Such cessation requires both a termination of the flow of negative *saṃskāra*s as well as a cultivation of "super *saṃskāra*." Yet again we see how the notion of *saṃskāra* represents a better "bridge" concept for understanding what is involved in Yoga, specifically the cessation of the turnings of thought.

In addition to practice and dispassion, Patañjali says that cessation can also result from various mental activities (I.17), as well as from faith (*śrāddha*), heroic energy (*vīrya)*, mindfulness (*smṛti)*, contemplative calm (*samādhi)*, and wisdom (*prajñā*)" (I.20). As Barbara Stoler Miller notes, this latter set of five powers accords with elements of early Buddhist practice.[25] Since Patañjali prefaces I.20 with "for others cessation of follows from..." we could argue that those "others" refer to such Buddhist communities in Patañjali's midst. Yet what is the status of this method for bringing about cessation? Are all types equally valid? Immediately before this aphorism, Patañjali says that beyond such cessation of turnings of thought only *saṃskāras* remain (I.18). This suggests that such methods do not *fully* bring about the cessation of thought (*citta-vṛtti*), for *saṃskāras* themselves have the potential to generate further thought. We might thereby conclude that such methods must be supplemented by practice: ultimately it is seedless contemplation (*nirbīja-samādhi*) that eradicates all *saṃskāras*. In this way, Patañjali affords some legitimacy to other religious practices while maintaining the supremacy of the practice detailed in his *sūtra*.

Patañjali suggests that devotional practice can bring about cessation of thought, specifically dedication (*praṇidhāna*) to the Lord (*īśvara*) of Yoga. He recognizes this Lord as separate from corruption (*kleśa*), fruits of action, and subliminal impressions (*saṃskāras*), and he then suggests that repetition of the syllable "*aum*," which he says encapsulates the sound of the Lord, will terminate any obstacles and distractions of the practitioner. We later see, however, that this method of bringing about cessation of thought is only preparatory. Barbara Stoler Miller claims that Patañjali's use of the term *praṇidhāna* suggests that it does not connote a worship of the Lord but instead committing to the discipline exemplified by the Lord.[26] Indeed, we may think of such dedication as that which calms the mind and prepares for an ultimate cessation of thought, just as the section on different types of tranquilities of thought (*citta-prasādhana*) describe means by which the practitioner can develop the equanimity for engaging in the seed-bearing and seedless contemplations. As Patañjali suggests, such tranquility of thought can result from the cultivation of friendship, compassion, joy, and impartiality (I.33), breath control (I.34), luminous or passionless thought (I.36, 37), or "when its foundation is knowledge from dreams and sleep" (I.38). This last aphorism signals that such calming is only preparatory, for sleep represents one of the types of *citta-vṛtti* mentioned by Patañjali at the beginning of the *sūtra* (I.6). Thus we see that such "cessations" are merely preliminary.

Indeed, these methods set the foundation for the practitioner to engage in those contemplations that bring about the ultimate cessation of *citta-vṛtti*, namely seed-bearing contemplation (*sabīja-samādhi*) and seedless contemplation (*nirbīja-samādhi*). Seed-bearing contemplation, which Patañjali equates with types of "contemplative poise" (*samāpatti*), follows the cessation of the generation of *citta-vṛtti* (I.41), though we will see that *citta-vṛtti* have not ceased completely, because *saṃskāra* still remain. Patañjali describes the state of seed-bearing contemplation as follows: "When memory is purified, then contemplative poise is free of conjecture, empty of its own identity, with the object alone shining forth" (I.43). Lest one think that the "purification of memory" implies an overcoming of all previous *saṃskāra*s, we should remember that we are still in "a state where only subliminal impressions remain from the practice of stopping thought" (I.18). Barbara Stoler Miller suggests that this reliance on memory enables the practitioner to realize the distinction between *puruṣa* and *prakṛti* in seed-bearing contemplation (Miller 42). Indeed, to suggest that the contemplation "bears seeds" means that there are still *saṃskāra*s that remain within one's consciousness from previous *citta-vṛtti,* even though no further *citta-vṛtti* are produced. Such *saṃskāra*s can bring about turnings of thought.

Only in seedless contemplation (*nirbīja-samādhi*) do such *saṃskāra*s cease completely. It is in this contemplation that wisdom (*prajñā*) emerges and "a subliminal impression generated by wisdom stops the formation of other impressions" (I.50). While Patañjali has mentioned previous cessations of *citta-vṛtti*, only this "super *saṃskāra*" ensures the full and complete cessation of *citta-vṛtti*. Patañjali writes, "When the turning of thought cease completely, even wisdom ceases, and contemplation bears no seeds" (I.51). If we reconsider Patañjali's definition of Yoga as bringing about the cessation of *citta-vṛtti,* we must recognize that only when the practitioner reaches this level of seedless contemplation does *citta-vṛtti* cease. For Patañjali, Yoga entails not only the preliminary cessation of turnings of thought, but also the ultimate cessation in seedless contemplation. That which brings about such seedless contemplation is the "super *saṃskāra*" that can erase even those *saṃskāra*s that remain in seed-bearing contemplation.

If we now consider the remaining parts of the *Yoga Sūtra,* we can assess the extent to which this revealed *puruṣa* participates in the material world. As suggested at the beginning of this paper, scholars have been particularly concerned with this issue. In his description of the practice of Yoga, Patañjali notes that its purpose is to "cultivate pure contemplation" (II.2). Yogic techniques serve to enable seed-bearing and seedless contemplation. Focusing on obstacles of ignorance, egoism, passion, hatred, and the will to live, Patañjali notes that the root of such obstacles are those subliminal intentions (*āśaya*) that are formed in actions (II.12) and the subliminal impressions (*saṃskāra*s) that perpetuate suffering (II.15). Only through meditation can the practitioner escape the effects of such obstacles (II.11)—especially, as we have previously noted, through the two types of contemplation.

Patañjali gives further evidence to suggest that the practitioner does not generate one "super *saṃskāra*" which stops the formation of all other *saṃskāra*s, but instead produces *multiple* "super *saṃskāras*" to ultimately bring about cessation of the turning of thoughts (*citta-vṛtti*). He writes: "The transformation of thought leading toward its own cessation is accompanied by moments of cessation, when subliminal impressions of mental distraction are overcome and those of cessation emerge in their place" (III.9). Thus Patañjali suggests that the discipline of generating *saṃskāras* from wisdom is a repeated endeavor. Here we get a better sense of what is entailed by the "maintenance of cessation" that we discussed previously vis-à-vis "practice" (*abhyāsa*).

Patañjali's description of such meditative contemplation might lead one to conclude that, in fact, the fully revealed *puruṣa* is quite removed from the world, yet I would argue that this would overlook Patañjali's claim that ultimately *puruṣa* stands as observer to *prakṛti*. Although he has previously stated that in its essence *prakṛti* exists only in relation to *puruṣa* (II.21), he later suggests that only through concentration on *prakṛti* can *puruṣa* eventually reveal itself: "From perfect discipline of the distinction between spirit as the subject of itself and the lucid quality of nature as a dependent object, one gains knowledge of the spirit" (III.35). Thus *prakṛti* plays a fundamental role in the revelation of *puruṣa*, for *puruṣa* itself cannot be the object of its own thought. Patañjali says ultimately, "absolute freedom occurs when the lucidity of material nature and spirit are in pure equilibrium" (III.55). This pure equilibrium consists of a *puruṣa* illuminating only the object (*prakṛti*) without mistakenly identifying itself with that object. In light of this type of equilibrium, it does not make sense to claim that *puruṣa* and *prakṛti* are absolutely distinct, for this would imply that there is no relation between them. Instead, in the final section of the *Yoga Sūtra,* Patañjali further explains how the *puruṣa* observes its last object, and how this observation differs from previous perception by the *yogin/yoginī.*

Lest one wonder whether *saṃskāra*s have disappeared completely from our discussion, the "super *saṃskāra*" returns in the final section of Patañjali's *Yoga Sūtra.* Here he begins the section by discussing karma—the ascetic practices in previous births that may have given generated powers of perfection (IV.1) and the overflow of material forces that occurs when people are reborn into different species (IV.2–3). He suggests that thought operates in a similar fashion for practitioners mired in deluded thinking: thoughts deriving from egoism (*asmitā*) fuel various activities of other thoughts (IV.4–5), and the turnings of thought (*citta-vṛtti*) continue unabated. As opposed to this cyclical dynamic of karma, Patañjali again recalls the power of the "super *saṃskāra*": "A thought born of meditation leaves no trace of subliminal intention" (IV.6).

This leads Patañjali to raise similar questions as contemporary scholars: if there are neither subliminal impressions (*saṃskāra*s), nor memory traces (*vāsana*s), nor subliminal intentions (*āśaya*s), then what is the karmic status of the *puruṣa*? In other words, does the *puruṣa* "act" at all? It is here, I would suggest, that the notion of *saṃskāra* can help clarify the way in which the *puruṣa* engages with the world after achieving such cessation. In this section of the *Yoga Sūtra*, Patañjali writes, "the action of a yogi is neither black nor white" (IV.7). *Yogis/yoginīs* indeed act, but their actions do not produce any traces. This is quite distinct from how karma functions, wherein *saṃskāra*s are sustained across birth, time, and place because the desires which support them are eternal (IV.9–10). *Yogis/yoginīs* are able to bring about the cessation of such karmic cycles by stopping the turnings of thought (*citta-vṛtti*), preventing further generation of subliminal impressions (*saṃskāra*s), and ultimately eradicating remaining subliminal impressions altogether through the generation of "super *saṃskāra*s." Patañjali writes, "Since the subliminal impressions are held together by the interdependence of cause and effect, when these cease to exist, the impressions also cease to exist" (IV.11). Patañjali describes an unraveling of cause and effect, such that actions no longer plant seeds that later come to fruition. Yet we must recall that the *yogi/yoginī* does act. Although the generation of "super *saṃskāra*s" does not leave a trace, the seedless contemplation demands disciplined action.

As if anticipating the next question about how the dynamic of karma affects *prakṛti*, Patañjali discusses material things and their essences over time. He suggests that *prakṛtic* things are real because they are unique throughout various transformations (IV.14). He writes, "Although an object remains constant, people's perceptions of it differ because they associate different thoughts with it" (IV.15). Here Patañjali underscores the fact that *prakṛti* remains unchanged over the course of time; it is only human perception of *prakṛti* that changes according to their mental associations. Without proper mental cultivation, people remain forever divorced from *prakṛti*, for their thoughts obscure its unique and unchanging reality. Just as *puruṣa* remains hidden behind *citta-vṛtti*, *prakṛti* is equally obscured. Thus Patañjali returns to the capacity of the *puruṣa* ultimately to master *citta-vṛtti*, such that it can perform its role as observer and perceive the observed *prakṛti* without obstruction. Patañjali suggests that both thought and its object cannot be comprehended simultaneously (IV.20)—in deluded consciousness, *citta-vṛtti* prevents a full appreciation of *prakṛti*. Again, he returns to the importance of the "super *saṃskāra*" that leaves no trace: "Awareness of its own intelligence occurs when thought assumes the form of the spirit through consciousness that leaves no trace" (IV.22).

Thus when we consider the relationship between *puruṣa* and *prakṛti*, we find that it is only through seedless contemplation (*nirbīja-samādhi*) that the *puruṣa* can observe *prakṛti* in its reality. *Prakṛti* becomes the single object of *puruṣa*, instead of thought, which takes everything as its object and leaves various traces and countless associations (IV.23–24). Although scholars generally fixate on *puruṣa* following cessation of thought, we should also attend to *prakṛti*, for it is equally occluded by such *citta-vṛtti*. Thus we must reconsider the aphorism: "One who sees the distinction between the lucid quality of nature and the observer ceases to cultivate a personal reality" (IV.25). We might not only interpret this as signaling the fact that the *puruṣa* does not misidentify with *prakṛti*, but also that the *puruṣa* appreciates the *prakṛti* in its own reality, which the *puruṣa* is now able to observe because of the cessation of turnings of thought (*citta-vṛtti*), which previously prompted various transformations of *prakṛti*. In other words, we can view the distinction between *puruṣa* and *prakṛti* as one of function: the *puruṣa* is the observer, and the *prakṛti* is the observed. Before seedless contemplation, the observed is not *prakṛti* but instead transformations generated by the turnings of thought. The *puruṣa* only appreciates the world in its reality when it is can see *prakṛti* as an observer views an object.

As if to underscore the significance of *saṃskāra*s, Patañjali concludes by drawing to mind those residual subliminal impressions that can prompt lapses in discrimination and the need to eliminate these traces through meditation (IV.27–28). Actions and forces of corruption cease to exist only when one engages in seedless contemplation, what he here calls "the essential cloud of pure contemplation" (IV.29). In this practice, not only is the *puruṣa* released from its bondage of *citta-vṛtti* (IV.31), but *prakṛti* as well ceases to be eclipsed by the sequence of transformations born of such thought (IV.32). Thus he concludes: "Freedom is a reversal of the evolutionary course of material things, which are empty of meaning for the spirit; it is also the power of consciousness in a state of true identity" (IV.34). Now we can reconsider the concept of *kaivalya* (aloneness, freedom) that Whicher proposes as a "bridge" concept. We should not only understand "freedom" as referring only to *puruṣa,* for it also applies to *prakṛti*. It is not *prakṛti* per se that is "empty of meaning" for *puruṣa*, but instead "the evolutionary course" that is deemed meaningless. As we have previously seen, *prakṛti* does not change over time – only those thoughts which the mind associates with *prakṛti* change. In fact, the *puruṣa* not only depends on *prakṛti* as a means of realizing itself, but ultimately as observer, it also relates to *prakṛti* as the object of its observation. Thus we might reconsider the status of the *yogin/yoginī* who is "*in* the world but is not defined by worldly existence."[27] In fact, the *yogin/yoginī* is indeed *in relation* to the world, as an observer to an observed, and both the *puruṣa* and *prakṛti* are freed from the previous bonds of *citta-vṛtti*. Yet the action of the *puruṣa* is not that of karmic cause and effect but of cultivating wisdom, which can generate those "super *saṃskāras*" that ultimately stop fruition altogether.

Thus we have seen that the notion of *saṃskāras* is essential for understanding the nature of Yoga as well as the relationship between *puruṣa* and *prakṛti* in Patañjali's *Yoga Sūtra.* His description of *saṃskāras* is innovative on two counts. First, he understands *saṃskāras* in an atypical way within the Hindu tradition, suggesting a psychological role as opposed to the typical understanding of *saṃskāras* as purificatory rites within Brāhmaṇical Indian society. Although he retains the Hindu notion of *saṃskāra* as something which marks a significant transformation, in the *Yoga Sūtra* this transformation is not social but psychological. As David Carpenter has suggested, this serves to synthesize traditional Brahmanical values (*saṃskāra* as means of purification through conformity to norms of dharma) and those of renouncers (*saṃskāras* of the inner life).[28] Patañjali's description of *saṃskāras* is also innovative because it suggests that there are not only *saṃskāras* that are past impressions playing a negative role in perpetuating *karma,* but also "super *saṃskāras*," which have soteriological significance, for only they can bring about complete cessation of *citta-vṛtti.* As I have suggested, *saṃskāra* would serve as a better "bridge" concept for understanding the philosophy and practice of Yoga as well as the relationship between *puruṣa* and *prakṛti* following cessation of *citta-vṛtti.* Responding to the controversy surrounding how engaged a *puruṣa* may be, we can state that the *puruṣa* is certainly more engaged in *prakṛti* following cessation of *citta-vṛtti*, for only then can it observe *prakṛti* without impediment.

Endnotes

[1] I originally wrote this paper in December of 2004; since then Ian Whicher has published an article entitled "The Liberating Role of *Saṃskāra* in classical Yoga" in the *Journal of Indian Philosophy* (2005) 33: 601–630. I refer the reader to that article; although there are points at which we differ, several of our arguments are similar.
[2] All quotations from the *Yoga Sūtra* are taken from Barbara Stoler Miller's translation, with the book (in Roman numerals) and verse number. *Yoga, Discipline of Freedom: The Yoga Sutra Attributed to Patanjali.* (Berkeley: University of California Press, 1995).
[3] Gerald James Larson and Ram Shankar Bhattacharya, eds. *Sāṃkhya: A Dualist Tradition in Indian Philosophy* (Princeton, N.J.: Princeton University Press, 19887), p. 23–27.
[4] Gerald James Larson. "Classical Yoga as Neo-Sāṃkhya: A Chapter in the History of Indian Philosophy," *Asiatische Studien* (1999), 53: 727.
[5] Christopher Key Chapple. "Yoga and the Luminous." In *Yoga: The Indian Tradition* (edited by Ian Whicher and David Carpenter, New York: RoutledgeCurzon 2003), pp. 83–96; Ian Whicher. "The integration of spirit (*puruṣa*) and matter (*prakṛti*) in the *Yoga Sūtra*," in *Yoga: The Indian Tradition* (edited by Ian Whicher and David Carpenter, New York: RoutledgeCurzon, 2003), pp. 51–69; Ian Whicher. *The Integrity of Yoga Darśana: A Reconsideration of Classical Yoga*. (Albany, NY: State University of New York Press, 1998).
[6] Ian Whicher. *The Integrity of Yoga Darśana: A Reconsideration of Classical Yoga* (Albany, NY: State University of New York Press, 1998), p. 65.
[7] Ibid., p. 185.
[8] Ibid., p. 210.
[9] Ibid., p. 272 and 281.
[10] Ibid., p. 274.

[11] Ian Whicher. "The integration of spirit (*puruṣa*) and matter (*prakṛti*) in the *Yoga Sūtra*," in *Yoga: The Indian Tradition* (edited by Ian Whicher and David Carpenter, New York: RoutledgeCurzon, 2003), pp. 51–69.
[12] Christopher Key Chapple. "Yoga and the Luminous." In *Yoga: The Indian Tradition* (edited by Ian Whicher and David Carpenter, New York: RoutledgeCurzon, 2003), pp. 83–96.
[13] Lloyd W. Pflueger. "Dueling with Dualism: Revisioning the paradox of puruṣa and prakṛti." In Ian Whicher and David Carpenter, eds. *Yoga: The Indian Tradition.* (New York: RoutledgeCurzon, 2003), pp. 70–82.
[14] Ian Whicher. *The Integrity of Yoga Darśana: A Reconsideration of Classical Yoga.* (Albany, NY: State University of New York Press, 1998) p. 292.
[15] Ibid.
[16] David Carpenter. "Practice Makes Perfect: The role of practice (abhyāsa) in Pātañjala yoga." In *Yoga: The Indian Tradition* (edited by Ian Whicher and David Carpenter, New York: RoutledgeCurzon, 2003), p. 35.
[17] Ibid., p. 41.
[18] Ibid., p. 42.
[19] Ibid., p. 43.
[20] Gavin Flood. *An Introduction to Hinduism* (Cambridge: Cambridge University Press, 1996), p. 64.
[21] Ibid., p. 201.
[22] Har Dayal. *The Bodhisattva Doctrine in Buddhist Sanskrit Literature* (Delhi: Motilal Banarsidass, 1970), p. 69-70.
[23] Louis De la Vallee Poussin. "Le Bouddhisme et le Yoga de Patañjali." *Mélanges Chinois et Bouddhiques* (1936–1937), 5: 230–231.
[24] George Feuerstein. *The Philosophy of Classical Yoga.* (Rochester, Vermont: Inner Traditions International, 1996), p. 78.

[25] Barbara Stoler Miller. *Yoga, Discipline of Freedom: The Yoga Sutra Attributed to Patanjali.* (Berkeley: University of California Press, 1995), p. 35.
[26] Ibid., p. 36.
[27] Ian Whicher. *The Integrity of Yoga Darśana: A Reconsideration of Classical Yoga* (Albany, NY: State University of New York Press, 1998), p. 292.
[28] David Carpenter. "Practice Makes Perfect: The role of practice (abhyāsa) in Pātañjala yoga." In Ian Whicher and David Carpenter, eds. *Yoga: The Indian Tradition* (New York: RoutledgeCurzon, 2003), p. 44.

Dharmamegha Samadhi and the Two Sides of *Kaivalya*: Toward a Yogic Theory of Culture

Alfred Collins

This essay will explore a liminal region lying between aspects of life that two orthodox Hindu schools of thought, Sāṃkhya and Yoga, appear to keep apart, namely existence in the world of flux and suffering and the experience of release from entanglement in psychophysical reality. I will use the rubric *duḥkha* for the realm of suffering, and also refer to it as "World One;" and *kaivalya* for the experience of release that I will also call "World Two." My aim will be to show how a paradoxical third realm, a "World Three" of culture, is opened between suffering and release (or between frustration and joy) through insight (*buddhi* or *jñāna*) that brings the worlds together precisely by sharply distinguishing them. Implicit in Yoga is a view of culture as being in essence the practice of living out enlightenment within the ordinary world, and so of transforming the everyday world of *duḥkha* into a realm of joy.

The opposition between *duḥkha* and *kaivalya* parallels the fundamental opposition in Sāṃkhya and Yoga between *prakṛti* (psychomaterial Nature) and *puruṣa* (consciousness or the conscious Spirit). Thus *kaivalya*, the aloneness or freedom sought at the end of the psychophysical process of Nature (*guṇa-pariṇāma-pravṛtti*), often seems to be a possibility only for the conscious spirit (*puruṣa*). Texts that suggest this interpretation include *Sāṃkhya Kārikā* (*SK)* XIX, *siddham sākṣitvam ... puruṣasya ... kaivalyam* ...., "It is established that *puruṣa* [as opposed to Nature or *prakṛti*] is a witness, possessed of aloneness," etc.; and *Yoga Sūtra* (*YS*) II. 25, *taddṛśeḥ kaivalyam*, "freedom of the seer from [the seen]."*

Even though *kaivalya* in essence belongs to *puruṣa*, it cannot be achieved through the action of *puruṣa* for the simple reason that *puruṣa* does not act. Only *prakṛti* can act "for *puruṣa*'s sake" (*puruṣārtha*). This is why both the *Sāṃkhya Kārikā* and *Yoga Sūtras* tell us in other places that *kaivalya* or its essential attributes are *not* limited to *puruṣa* but are possibilities for the highest evolute of Nature, namely the mind whose faculty of discernment (*buddhi*) is purified (*sattvikā*, *SK* 23). For instance, at *SK* 64 the insight of a *buddhi* that discriminates the mind from the conscious spirit is called *kevala*, "solitary" or "free." *YS* IV.26 (*tadā viveka-nimnam kaivalya-prāgbhāram cittam)* states "Then, deep in discrimination, thought tends toward *kaivalya*" (or "gravitates toward" *kaivalya* as rendered by Barbara Stoler Miller). Most explicitly, the last verse of the *YS* (IV.34) gives two definitions of *kaivalya*, one from the point of view of purusa, the other from the perspective of *prakṛti* when her psychomaterial constituents or aspects have "flowed backwards" (*pratiprasava*) and ceased their endless efforts to give *puruṣa* pleasure (*bhoga*) and release (*kaivalya*). Interpretation of this verse, and especially the phrase *puruṣārtha-śunya* will be a central concern of my paper.

Barbara Stoler Miller's translation of the verse clearly shows that it names two distinct meanings of *kaivalya*, although I think she misconstrues the first sense:

> Freedom [*kaivalya*] is a reversal of the evolutionary course of material things, which are empty of meaning

---

* Textual citations from the *Yoga Sūtras* are taken from J.W. Hauer, *Der Yoga*, Verlag Bruno Martin, 1983. Citations from the *Sāṃkhya Kārikā* are from James Larson, *Classical Sāṃkhya*, Motilal Banarsidass, 1969.

> for the spirit; it is also the power of consciousness in a state of true identity.[1]

The first words of the verse, a phrase in apposition to the word *guṇas* (the constitutents of psychophysical materiality) are *puruṣārtha-śūnya*, which Miller translates as "empty of meaning for the spirit." The problem with this rendering is that it hides the significance of the crucial word *puruṣārtha*, "for *puruṣa's* sake," which the last substantive verse of the *Sāmkhya Kārikā* (69) names *guhya-jñāna* ("secret wisdom"). I suggest that the idea of *puruṣārtha*, the assertion that Nature (*prakṛti*) acts to give enjoyment and release to the conscious spirit, is the central thought in both the *Sāmkhya Kārikā* and the *Yoga Sūtra*. The idea, simply, is that everything that happens—or more precisely everything that *acts or behaves* in our essentially "personal" psychophysical world—does so in order to give enjoyment and release to the conscious spirit which is the only real self in each person and indeed in all living things, "from Brahmā to a blade of grass" (*SK* 54). Thus *YS* II.18:

> *bhoga-apavarga-artham dṛśyaṃ*
>
> The aims of the phenomenal world are the enjoyment and release (of *puruṣa*).

There is, however, a second self founded on ignorance, called *asmītā* in the *YS* and *ahaṃkāra* in the *SK*, that lives within World One, the realm of *duḥkha*. Evidently there is a contradiction: the relationship of these two selves, *puruṣa* and *ahaṃkāra*, must be, following the principle of *puruṣārtha*, that *ahaṃkāra* acts to give *puruṣa* enjoyment and liberation, yet the second verse of the *SK* and other places in both texts attest that life is essentially suffering. Apparently this world of psyche and matter that lives to unfold a little story told for *puruṣa's* enjoyment and edification has gotten stuck in a complex, evolving Scheherazade tale of many thousand nights about the adventures of a creature made in *puruṣa's* image but of a wholly different nature. *Ahaṃkāra*, as its synonym *abhimāna* suggests, lives for its own sake rather than *puruṣa's*—or at least it *thinks* it does, and that is enough to keep the story going. Paradoxically, the *denouement* of our tale requires that the *ahaṃkāra* turn back the on-rolling history that it is part of—indeed, of which it is both the protagonist and the antagonist—and do its duty, its *dharma*, of serving *puruṣa*.

Thankfully, the *ahamkāra*—or rather the subtle body or *liṅga-śarīra* into which it evolves, can be seen through and overcome by the intellect (*buddhi*), which lies above it in the hierarchical emanation-structure of the world. The texts make clear that there are two kinds of *buddhi*, one that is afflicted (*kliṣṭa*) and flows forward, and another, smaller category that flows backwards and is not afflicted (*akliṣṭa*). The afflicted process of living in the world, though it works for *puruṣa's* enjoyment and liberation as much as does the non-afflicted, leads to entanglements in the *ahaṃkāra*. Turning the process back (*nivṛtti, pratiprasava*) leads out of life's mess and towards *kaivalya*. Afflicted *buddhi* constitutes ordinary life, the realm of suffering, while non-afflicted *buddhi* forms the basis for enlightenment and culture. In turning back the clock of *prakṛti's* evolution, culture—and Yoga and philosophy are central parts of culture—moves towards the early state (imagine the pre-"big bang" universe) when the evolutionary flow had not yet evolved into goal-seeking manifestation. This state, called *avyakta*, the "unmanifest" or "indeterminate," was in the beginning just a condition of the three *guṇas* balanced in equilibrium. Approached through the insight (*buddhi*) that lies at the heart of culture, this "indeterminate" stage can be regained, and develops into a golden age of culture, a Vrindivan (the blessed city). Culture is innately conservative because it involves this turning back the clock of *pariṇāma*, yet the new world that arises, though it reprises the oldest world, has a different flavor, exactly as the *turīya* state described in the *Māṇḍukya Upaniṣad* (the "fourth" state of consciousness that lies beyond the three ordinary realms of waking, dream, and dreamless sleep), paradoxically reinterprets deep sleep. Culture lives on the edge of enlightenment, always oriented toward it but equally always aware that, as part of the psychomaterial world, it is "not" that.

Let us follow the process of life as the *SK* and *YS* lay it out for us, in its two possibilities of ignorance and insight. In the beginning, or *in illo tempore* as Mircea Eliade put it, there was *puruṣa* and an indefinite, balanced state of the three constituents of materiality: intelligibility, passionate activity, and lethargy (*sattva, rajas, tamas*). In order to give *puruṣa* enjoyment and release, prakṛti began the long process of unfolding herself (*pariṇāma*). Early in this process, her insight into her purpose (which always is *puruṣārtha*) became afflicted or deluded. This delusion concretized into the principle of egoity, termed *ahaṃkāra* or *asmitā*. Subsequently, most actions have been motivated (or seem to be motivated) by self-interest. This is World One, the realm of *duḥkha*.

Bhukti and *mokṣa* are identical goals, or at least their essence is the same. The situation is parallel to the root identity of Freud's drive reduction and death instinct.[2] Drive reduction, like the state of a person quenching a desire in the *YS/SK*, focuses on the dying down of the *pariṇāma* (this can be momentary or permanent), whereas the death instinct and *mokṣa* (*nirvāṇa*) look at the goal as it is in itself and not from the viewpoint of the suffering individual.[†] This is why *kaivalya* has two sides. One side is the dying down of the deluded process of *pariṇāma* into a state of quietness and insight, the other is the consciousness principle (*puruṣa*) that *buddhi* intuits just beyond its horizon, in the form or figure of its own non-being (*nāsmi*, *SK* 64).

---

† I have addressed the question of the Freudian death instinct as a transcendental principle at greater length in a later paper (Collins, 2008).

The essence of this ultimate intuition of the *buddhi*, which is the nature of *buddhi* in *kaivalya,* is stated most clearly in *SK* 64 and *YS* IV.34, at the ends of the two texts. It is also expressed through a series of metaphors and—as I hope to show—more generally in the form of genuine culture. Yoga's theory of culture is *critical*, in a way like that of Theodor Adorno and other forms of Marxism. Some forms of culture are *true* and lead towards freedom and enlightenment; other forms, which I would characterize as *false* culture (or false consciousness), lead to bondage and suffering. Let us first look at the two verses, then at the metaphors, and finally at several examples of Western culture and culture critique that may embody World Three's simultaneous affirmation and denial of self.

SK 64

evam tattvābhyāsān nā 'smi
na me nā 'ham ity apariśeṣam
aviparyayād viśuddham
kevalam utpadyate jñānam

From practice [of the stopping or reversing of the flow of *prakṛti's pariṇāma*] there arises the insight that 'I am not,' 'I possess nothing,' and 'there is no I [in me].' This insight is complete, pure and solitary because there is no illusion in it.

YS 4.34

purusārtha-śunyānām guṇānāṃ pratiprasavaḥ
kaivalyam svarupa-pratiṣṭhā vā citiśaktir iti

Freedom (*kaivalya*) is the flowing backwards of the *guṇas* whose work for the sake of *puruṣa* has become (their own) emptiness. *Kaivalya* is also (*vā*) the power of consciousness in its own reality.

Freedom, then, has two sides. From the point of view of *puruṣa,* freedom is the placement of *citiśakti*, the power of consciousness, in itself. From the point of view of *prakṛti*, it is a timeless nay-saying of the *ahaṃkāra* self, a *nāsmi* that empties the purposes for which the *ahaṃkāra* has striven of all meaning because they are being achieved in this moment. Neither side of *kaivalya* is static: *puruṣa* is identified as having *śakti*, and *prakṛti* is engaged in *nāsmi*-saying in its practice of *pratiprasava* or *nivṛtti*. The flowing-back of *samādhi* is simultaneous with an independent shining-forth of *puruṣa*. *Prakṛti* in *kaivalya* lives in a state where insight and quietude (*vijñāna* and *samādhi*) have become one, so that discrimination is constant. It is as if *prakṛti* lives on the edge of a black hole—*puruṣa*—into which it perpetually flows while paradoxically remaining separate.

The nature of *prakṛti's* enigmatic existence in *kaivalya*, and the intimate connection that *kaivalya* establishes between *prakṛti* and *puruṣa*, are suggested by several metaphors in the *SK* and *YS* which include 1. the syllable *aum*, 2. the Lord of Yoga (*yogeśvara*), 3. a spinning potter's wheel, and 4. a rain cloud dense with moisture. All these figures suggest a kind of self-contained or circular motion, which is most evident in the image of a potter's wheel found at *SK* 67 (*cakra-bhrami*). The circularity of *aum* is evident elsewhere in its representation of the beginning, middle, and end of the world (a-u-m), as in the *Māṇḍukhya Upaniṣad.* The *YS* identifies the Lord of Yoga with the sound *aum*, and tells us that this Lord is the guru of the ancient seers (*YS* I.26) and the source of universal wisdom (*sarva-jña*). Most interesting is that this Lord of Yoga is identified as a "sort of *puruṣa* " (*puruṣa-viśeṣa*), *YS* I.24. Clearly the categories are being brought together here, as it is ordinarily the *distinction between puruṣa* and *buddhi*, even the most *jñāna*-infused *buddhi*, that leads to *kaivalya*—and yet here the special *puruṣa* called Lord of Yoga directly communicates the *jñāna*, as a guru to a disciple.

The point, I think, is that here the aim (*artha*) of *prakṛti* is not separate from its nature. There is no projection of *puruṣa's* release or enjoyment into the future but rather a here-and-now experience (*bhāvana*) of it. Hence the repetition of the syllable *aum* leads to the experience of the aim which lies within it. Thus *YS* I.28

taj-japas tad-artha-bhāvanam

"Chanting the syllable *aum* is the experience (*bhāvana)* of its meaning."

There is no separation between the word and meaning, no seeking after something unattained, a *puruṣārtha* located in the future. *Puruṣārtha* is still the central idea, but it is now something timelessly *found* rather than something to be *sought*. There is a sense here of finality or fulfillment, no longer of going somewhere or seeking a state not yet attained, but also not a dead or rigid stasis. As the sculptor Anish Kapoor says of his own cultural productions when they are "well made," this is a "condition that seems to be abidingly static and at the same time dynamic. It's hard to name but it's a condition that I just know exists when it happens. . . ."[3]

The idea of celebrating or enacting—to use a postmodern term, we might say *performing*—the recognition of *puruṣārtha* is also present in the image of the raincloud, *dharmamegha-samādhi*, representing the highest stage of meditation in the *Yoga Sutra*. This state implies an easy, disinterested flow of discriminative insight (*viveka*) that is not oriented toward gain in the future (*akusīda*, *YS* 4.29). The same image of *dharmamegha* is also found in Buddhism, as discussed by Klostermaier.[4] There, it represents the state of the *bodhisattva* who is ready to enter nirvāṇa but no longer strives for it, instead spontaneously bestowing the blessing of insight on others. This in-between moment develops in Buddhist mythology into the various Buddha realms and kingdoms of jewels—ideal forms of culture, as in Hinduism it evolves into such utopias as Vrindavan, the "world between the guru and the disciple"[5], the tantric kingdoms of Kasmir and Nepal (real or imagined), and so forth. In a word, *dharmamegha samādhi*, like the life of the *bodhisattva*, *is* authentic culture. The essence of culture is its gift of insight into the true nature of *puruṣa* and *prakṛti*.

In the West also, such moments of insight are not rare. I find an illustration, characteristically tinged with fear, in Leonard Cohen's song about Joan of Arc, where a dialogue between Joan and the fire in which she was burned is imagined as a wedding between a principle of consciousness and one of materiality.

> she climbed inside
> To be his one, to be his only bride.
> And deep into his fiery heart
> He took the dust of Joan of Arc,
> ....
> And then she clearly understood
> If he was fire, oh then she must be wood.
> ....
> Myself I long for love and light,
> But must it come so cruel, and oh so bright?

To recognize consciousness, *puruṣa*, is to be burned thoroughly, to say and fully realize that one's essence is *nāsmi, na me, nāham*. Similar to Leonard Cohen's song, sexual union between *puruṣa* and *prakṛti* is implied in the *SK* when *prakṛti* dances before *puruṣa*. Already a goddess at the time the text was written, *prakṛti*'s femininity is clearly implicit throughout the *SK*. Similarly, *puruṣa* evokes the Cosmic Man he formerly was[6], and he is called *puṃs* in the *SK*, which underlines his maleness. The resonances of the *puruṣa/prakṛti* relationship in later tantra and bhakti also reflect the sexual *hieros gamos* of their tie. Joan recognizes in her death the fire's love and light; and while the singer suffers the burning heat and blinding light involved in surrender to the consciousness principle, we are also shown Joan's epiphany of understanding that "he was fire" and "she must be wood." Not quite wood, even, for in the moment Joan's body is burned away she is identified as "ash." This ashen Joan is quite parallel to the *nāsmi*-singing *prakṛti* of the *Sāṃkhya Kārikā* whose *ahaṃkāra*-based personality likewise dissolves in *jñāna*.

Other illustrations of the Third World of culture are suggested in the work of the Indian-British sculptor Anish Kapoor and the Indian culture critic Homi Bhabha. Kapoor's recent mirrored pieces could almost be considered reflective analogues of a refractive metaphor central in Sāṃkhya, that of a piece of rock crystal within which the image of a flower (actually placed nearby) seems to float. The sense of this image is that the flower appears to be enmeshed in the crystal but really is outside it, just as experience (*prakṛti*) seems to be encased in consciousness (*puruṣa*) but really does not touch it. Kapoor's work, in which the cityscape of the Chicago Loop seems to float like a "world of dreams" is described by Bhabha as "making emptiness." It could equally be characterized as "emptying the made." It is about the edge between the object and awareness of it, and it carries us from one world to the other—from World One to World Two—while itself being part of a third realm of culture not reducible to either of the others.

Homi Bhabha, from a postmodern and postcolonialist stance on the Mumbai—Harvard axis, speaks of a third world in his concept of "splitting" between a colonialist realm of lies and an underlying reality that the colonial regime denies or hides.[7] The third world of splitting involves a particular kind of strategy for moving from authoritarian deceits to the suppressed reality. Briefly, Bhabha finds his third world, that of liberating culture, in "an enunciatory space, where the work of signification voids the act of meaning...." Notice how similar this is to Sāṃkhyā where the *prakṛti*'s work of signifying *puruṣa* voids that act in the very saying of "not I." The situation Bhabha is studying occurs within discourse—not only colonial discourse—when "the resistance to authority, the subversion of hegemony" operates with a strategy of "disarticulating the voice of authority at that point of splitting," when the subaltern colonized or social underclass uses the language of oppression against that very oppression. This move is formally identical to the strategy employed by Sāṃkhya/Yoga when it teaches the use of *prakṛti* against *prakṛti*, or "non-afflicted" *buddhi* against "afflicted" *buddhi*. The proportion is: *ahamkāra-buddhi* (in Sāṃkhya) is to Bhabha's "voice of authority" as *kaivalya-buddhi* is to "competing discourses of emancipation or equality" and the forms of "identity and agency" (i.e., forms of true culture) that arise from them. *Ahaṃkāra* and the

Anish Kapoor's Cloud Gate

voice of authority are based on ego motivations; *kaivalya* and the discourse of emancipation touch a state beyond ego. The experience of *puruṣārtha* takes emancipation to depths that Bhabha may not recognize but which his thought approaches.

In a highly interesting critical essay by Homi Bhabha about the work of Anish Kapoor[8] Bhabha again arrives at a position that is quite near that of Sāṃkhya and Yoga, and in he fact uses terms taken by Kapoor from classical Indian thought that are related to the ideas of our texts, namely *rūpa* and *svayambhū* ("man made"/"self made" in Kapoor's terms). Clearly these words are related to *prakṛti* (*rūpa*) and *puruṣa* (*svayambhū*), and name the human- (or *ahaṃkāra-*) shaped quality of the world on one hand, and its self-existence on the other. The central argument of this essay, and I think the center of Sāṃkhya-Yoga, is that neither *puruṣa* nor *prakṛti* can really exist "in itself" without the other, or rather that *prakṛti* always exists for *puruṣa* and *puruṣa* exists in itself only through *prakṛti.* There is always *puruṣārtha* to direct *prakṛti*'s action, and *prakṛti* for *puruṣa* to see. Bhabha and Kapoor understand something similar, and find a back-and-forth movement between *rūpa* and *svayambhū* that establishes a third thing, an ambiguous and flickering reality that Kapoor names "Ghost" in one of his works. Kapoor proposes the useful term "truly made" to name the products of genuine culture, which occur "only when the material and the non-material tangentially touch."[9] As Kapoor says, a "thing exists in the world because it has mythological, psychological, and philosophical coherence. That is when a thing is truly made."[10]

Yogic culture, then, which I believe we may fairly identify with the true culture and culture theory of India, consists of those "truly made" objects and ideas that inhabit their own *śūnya*, (emptiness), their own *nāsmi* (saying "not-I"), and in this circular dynamism of being/nonbeing point us in the direction—ironically, the direction we were already headed without knowing it—of *puruṣārtha*. And this is true not only for Indian culture. Contemporary Western art, music, and criticism suggest the same thing, though perhaps without the full experience, earned through *dhāraṇā, dhyāna*, and *samādhi,* of the *puruṣa* that shines where we are not. This culture theory is a gift that India's ancient thought still offers the world, one that, as we will see, can hold ecological significance.

Endnotes

[1] Barbara S. Miller. *Yoga Discipline of Freedom: The Yoga Sutra Attributed to Patanjali* (University of California Press, 1995), p. 83.
[2] J. Laplancheand J.-P. Pontalis. *The Language of Psychoanalysis* (New York: W.W. Norton, 1973), p. 102.
[3] Homi K. Bhabha. "Anish Kapoor: Making emptiness" in *Anish Kapoor*, (Hayward Gallery and University of California Press. Bhabha, 1998), p. 28.
[4] K. Klostermaier. "Dharmamegha Samadhi," in *Philosophy East and West* (University of Hawaii, 1986), 36, 3:253–262.
[5] Sri Adwayananda (K. Padmanabha Menon). *Radhamadhavam: Explanation of the Spiritual Poem* (Austin: Advaita Press, 1986).
[6] J. A. B. Van Buitenen. "Studies in Sāṃkhya II". *Journal of the American Oriental Society,* 77 (1957), pp. 15–25.
[7] J. W. T. Mitchell, translator. "Interview with culture critic Homi Bhabha," *ArtForum*, 33 (1995), p. 80–84.
[8] Anish Kapoor, 1998.
[9] Bhabha, p. 18.
[10] Ibid., 18.

Toward a Yoga-Inspired Environmental Ethics

Christopher Key Chapple

The prior three chapters argue that the Yoga tradition of Patañjali takes the manifest world seriously, not only in terms of overcoming its attendant sorrows, but as a fundamental and necessary condition for all human experience, including the experience of liberation. Ian Whicher proclaims here and elsewhere that *kaivalyam* does not require a complete dismissal of the world but suggests the possibility of a continued yet auspicious, conscious encounter with the realm of the manifest after liberation. Beverly Foulks provides a detailed assessment of what she labels the "super *saṃskāra*s" generated in states of seedless (*nirbīja*) *samādhi* suggesting that actions informed by this experience hold redemptive power. Her discussions of karma in its purified (*akliṣṭa*) state posit a psychological bridge between *puruṣa* (consciousness) and *prakṛti* (the manifestation of action). Alfred Collins probes the image of *dharma megha samādhi*, which he renders as a "rain cloud dense with moisture," suggesting that it offers a metaphor for engaging the world in "an easy, disinterested flow of discriminative insight," resulting in "authentic culture." In this chapter, I will argue that Whicher's affirmation of worldly engagement even after *kaivalyam*, Foulk's articulation of the efficacy of *nirbīja samādhi* as a psychological bridge between transcendence and involvement, and Collins' assertion of Yoga's "circular dynamism" find support in the ethics of Yoga, for which I will suggest an ecological application.[1]

The *Yoga Sūtra* of Patañjali outlines an eightfold practice to ascend toward the state of ongoing spiritual discernment (*viveka khyāti*). The underlying philosophy of Yoga places great value on feeling the connection between one's self and the larger world of nature. This continuity becomes celebrated in the term *samādhi*, the goal of Yoga, which describes an experience of non-difference between oneself, one's sensory and mental processes, and the world. As described by Patañjali, the practitioner of Yoga becomes like a clear jewel, with "unity among grasper, grasping, and grasped" (*Yoga Sūtra* I:41). This state of consciousness allows one to melt into one's surrounding and in the process diminish and eventually reverse past tendencies (*saṃskāra*) bringing one to a state of clarity and immediacy. Whicher, as mentioned above, focuses on *kaivalyam*; Foulks on *nirbīja samādhi,* and Collins on *dharma megha samādhi* as holding resonance with environmental themes.

The eight practices of Yoga identified by Patañjali can be seen in light of environmental ethics. The beginning stage of Yoga includes five ethical practices (*yama*), held in common with the Jain tradition. First and foremost, Patañjali discusses nonviolence (*ahiṃsā*), which entails not harming any living being by thought, action, or assent to harmfulness. This precept advocates the protection of all forms of life, and certainly can be applied to cultivating an attitude of respect toward individual creatures as well as ecosystems. Nonviolence or *ahiṃsā* forms the foundation for developing an inclusive ecological ethic. By attempting to do no harm to any life form, one must take into account the interconnections of all life forms. If harm is done to the air, earth, or water, effects will be found in those beings that dwell therein. The diminishment or sacrifice of one species will affect other species as well, making life more difficult. The protection of one life form may very likely allow others to flourish as well. A conscious practice of nonviolence or *ahiṃsā* will provide an ongoing point of conscience when making decisions that have environmental impacts, including the choice of one's food, the choice of one's car, and even the choice of one's Yoga mat.

To support the discipline of nonviolence, Patañjali includes four additional vows. Truthfulness (*satya*) can be used to inspire acknowledgement of wrongdoing to the living realm. Not stealing (*asteya*) can be applied to remedy the imbalance of resource consumption in modern times. Sexual restraint (*brahmacarya*) can be used as a corrective to the crass commercialization of sex as well as for population control. Nonpossession (*aparigraha*) allows one to minimize the greed and hoarding that has plundered the planet. By restricting one's ownership of things, one is able to release attachment from external objects. The market driven economy relies on constant growth in the consumer sector. Advertising, seductive shops, and the bombardment of media from the internet to television to magazines draw a person out from his or her core into a world of false and ephemeral images. Mahatma Gandhi, though he certainly supported and promoted the development of industrial technology, advocated a locally-based village economy. Each village would produce its own food, grow its own cotton, spin its own thread, and weave its own clothes. Although an utterly local economy is not very practical in today's urbanized, globalized business environment, buying strategically and sparingly can help contribute to one's own health and the health of others.

These five practices entail holding back, disciplining oneself, saying no to such behaviors as violence, lying, stealing, lust and possessiveness. The second stage of Patañjali's Yoga seeks to cultivate positive behaviors that can similarly be interpreted through the prism of heightened ecological awareness. Five practices are listed. Purity (*śauca*) can be seen in terms of keeping one's body, thoughts, and intentions clean in regard to one's surroundings. Contentment (*santoṣa*) encourages a philosophy of accepting what is "enough" and not striving to gather more than one truly needs. Austerity (*tapas*) entails putting oneself in difficult situations for the purposes of purification and the building of strong character. Self study (*svādhyāya*) generally entails reading and reflecting on philosophical texts and in the case of environmental applications might include reading the nature poets. Devotion to god (*īśvara praṇidhāna*) for an environmentalist might encourage regular forays into the wilderness to feel that important connection with the awe that nature inspires. Each of these serves as a touchstone for self-exploration and appreciation of one's place within the world.

The third phase of Patañjali's eightfold system, the practice of Yoga postures (*āsana*) receives relatively scant mention in the *Yoga Sūtra.* Patañjali states that the purpose of performing the physical exercise of Yoga is to gain "steadiness and ease, resulting in relaxation of effort and endless unity" (*YS* II:46–47). In later centuries, this aspect of Yoga was expanded by later writers, who draw extensive parallels between the practice of physical Yoga and the ability to see one's relationship with the animal realm. In the later Yoga texts, animals play an important role. Many postures (*āsanas*) carry the names of animals. The *Haṭha Yoga Pradipika*, written by Svatmarama in the 15th century, lists several poses named for animals. Some examples are the Cow Head's pose (*Gomukhāsana*) [HYP 20], the Tortoise Pose (*Kūrmāsana*) [HYP 24], the Rooster Pose (*Kukkutāsana*) [25], the Peacock Pose (*Mayūrāsana*) [32], and the Lion's Pose (*Siṃhāsana*) [52-54]. Additionally, later Yoga manuals such as the *Gheraṇḍa Saṃhitā* include several additional poses named for animals, including the Serpent Pose (*Nāgāsana*) the Rabbit Pose (*Śaśāsana*) the Cobra Pose (*Bhujāṅgāsana*), the Locust Pose (*Śalabhāsana*), the Crow Pose (*Bakāsana*), the Eagle Pose (*Garuḍāsana*), the Frog Pose (*Maṇḍukāsana*), and the Scorpion Pose (*Vriśchikāsana*), to name a few.

Following the mastery of the physical realm through Yoga postures, one reaches the capacity to effectively control the breath (*prāṇāyāma*), the fourth phase of *Yoga.* As noted earlier, the breath plays an important role in the philosophy of the Upaniṣads, and in the *Yoga Sūtra* the mastery of the inbreath and outbreath leads to "dissolving the covering of light" (*YS* II:52). The *Hatha Yoga*

Pradīpikā and the *Gheraṇḍa Saṃhitā* describe intricate techniques for manipulating the breath. Through this process, one reaches into the core of one's life force, sees the relationship between breathing and thinking, and cultivates an inwardness and stability, leading to Patañjali's fifth phase, the command of the senses (*pratyāhāra*). This ability to draw one's energy into oneself opens one to the higher "inner" practices of Yoga: concentration, meditation, and *Samādhi*, collectively known as *saṃyama*. Construed through an ecological prism, the inner work from controlling the breath to *Samādhi* can be seen as enhancing one's sensitivity to nature, an increase in empathy, and a willingness to stand to protect the beauty of the earth. In a sense, the culmination of Yoga leads one to the very beginning point of nonviolence, a sense that no harm must be allowed.

The beginning of this inner three-fold process requires sustained exercises of concentration (*dhāraṇā*). A standard concentration practice entails attention given first to the great elements (*mahābhūtas*), then to the sensory operations (*tanmātras*), the sense and action organs (*buddhīndriyas* and *karmendriyas*) and finally to the three fold operation of the mind (*manas, ahaṃkāra, buddhi*). By concentrating on the earth (*pṛthivī*) one gains a sense of groundedness and a heightened sense of fragrance. By reflecting on water (*jal*), one develops familiarity with fluidity and sensitivity to the vehicle of taste. Through attention to light and heat (*tejas, agni*), one arrives at a deep appreciation for the ability to see. Awareness of the breath and wind (*prāṇa, vāyu*) brings a sense of quiet and tactile receptivity. All these specific manifestations occur within the context of space (*ākāśā*), the womb or container of all that can be perceived or heard.

Intimacy with the sensory process allows one to maintain focus on the operations of the mind. Thoughts (*citta-vṛtti*) generated in the mind lead one to question and investigate the source of one's identity and ego (*ahaṃkāra*). Probing more deeply into the constituent parts of one's personality, one begins to uncover the maze and mire of karmic accretions housed in the deep memory structures (*buddhi*), lightened and released gently through reflective and meditative processes. However, in order for any of these purifications to arise, an intimate familiarity with the body and collection of habits must occur, an intimacy that takes place through an understanding of time and place. Yoga enables a person to embrace and understand the close connection between the body and the world. By understanding each, one attains a state of clarity.

Beverly Foulks has characterized the purifying aspects of Yoga as generating "super *saṃskāra*s." Patañjali provides a description of the efficacy of purposeful, ethical action to explain the benefits of cultivating a lifestyle grounded in the restraints or precepts (*yama*) and the positive observances (*niyama*) described earlier in this chapter. According to Patañjali, in any situation where thought and action lead to bondage (*vitarka bādhane*), one should cultivate the opposite (*pratipakṣa bhāvana*, *YS* II:33). He further delineates the nature of these thoughts and actions, specifying that they are rooted in violence, lust, anger, and delusion, that they can be done directly or caused by others, and that they have varying degrees of severity. Regardless of their context or intensity, these thoughts and actions lead to "endless suffering and ignorance" and must be countered through the cultivation of opposite behavior.[2]

This insightful description of how negativity cascades into a downward spiral invites an ecological interpretation and application. Violence to the environment can be seen in the increase of health issues affecting both humans and species of plants and animals, often resulting in death and extinction. Violence can also be seen in the growingly evident effects of global warming. Due to ignorance, the human pursuit of happiness through a manipulation of the environment and over-exploitation of is resources has led to unimagined deleterious consequences. Lust for money, comfort, and power as well as anger whenever commercial forces are thwarted in their attempt to "deliver the goods" result in delusion, disaffection and even war. Some of the effects can be easily discerned throughout human history, such as the tragedy of strip mining for coal and the wholesale destruction of mountains for resource extraction. Other effects are more subtle, seen in rates of learning disabilities and depression. According to Patañjali, the remedy must be found through the application of the human will. By applying such as techniques nonviolence, honesty, restraint, contentment, and devotion, one can move away from the impulses of greed and delusion and create thoughts and activities that lead to purification and ultimately freedom.

Vyasa commented upon the human predicament with an apt analogy. Without the cultivation of proper behavior informed by the ethics of Yoga, humans will return to actions riddled by greed and delusion "like a dog returns to eat its own vomit."[3] Vyasa also invokes images of the afterlife, proclaiming that one who causes injury to animals or humans (and by extension to nature herself) will "experience pain in hells and in the bodies of animals and of departed spirits in other forms" in future rebirths.[4] In our circumstance of ecological devastation, we need not wait for future births. We feel and see the effects of eco-violence in the present. Furthermore, a common refrain in regard to future life focuses not necessarily on our own rebirth but on the legacy left to our progeny. Thinking about future generations can help stimulate one to cultivate corrective behavior in regard to pollution and general lifestyle issues.

From an ecological perspective, the practice of Yoga can prove beneficial. Through Yoga one can begin to see the importance of the food we eat in constructing our bodies. One can find a calmness of mind through which to appreciate the stunning beauty of landscape and sunset and sunrise. Through Yoga, one can understand that all things within the universe rely on the creative expression of the five great elements and that we gain access to all experience and all knowledge through our own sensuality and intuition. The practice of Yoga provides rich resources for persons to reconnect with the body and with the world. In its various manifestations, Yoga includes practices and philosophical positions that seem in alignment with values espoused by modern ecologists and environmental ethicists. Through the insights and applications of Yoga, one can begin to live with the sensitivity, sensibility, and frugality required to uphold the dignity of life, stemming from a vision of the interconnectedness of all things.

Endnotes

[1] Earlier discussions of applied environmental ethics from the perspective of classical Yoga can be found in Chapple, "Toward an Indigenous Indian Environmentalism" in Nelson, *Purifying the Earthly Body of God*, pp. 29–31; *Nonviolence to Animals, Earth, and Self in Asian Traditions*, pp. 53–57; "Yoga" in the *Encyclopedia of Religion, Ecology, and Culture*; in *EcoYoga* by Henryk Skolimowski; and in the writings of Laura Cornell, including pp. xxxx in this volume. Yoga scholars Georg and Brenda Feuerstein have published a book titled *Green Yoga* which also explores these themes (Vancouver: Traditional Yoga Studies, 2008).

[2] *vitarkā hiṃsādayah kṛta-kārita-anumoditā lobha-krodha-moha-pūrvakā mṛdu-madhya-adhimātrā duḥkha-ajñāna-ānanta-phalā iti pratipakṣa bhāvanam* (*Yoga Sūtra* II:34).

[3] Commentary on *YS* II:33. See James Haughton Woods, *The Yoga System of Patanjali* (Cambridge: Harvard University Press, 1914), p. 183.

[4] Woods, p. 184.

The Disharmony of Interdependence: Sāṃkhya-Yoga and Ecology

Knut A. Jacobsen

The Yoga of Deep Ecology

Traditions from Hinduism, especially Hindu traditions that emphasize renunciation, have had a significant influence on several global environmental movements. In this paper I will first discuss the influences of Yoga traditions on one of these movements, the Deep Ecology movement, and thereafter I present some Sāṃkhya-Yoga views of the natural environment.

The Deep Ecology movement founded by the Norwegian philosopher Arne Naess (1912–2009) was deeply influenced by Hindu ascetic traditions and the Indian monistic philosophy of Vedānta.[1] Arne Naess before he started to publish his deep ecology ideas was well known within philosophy for his books on Mahatma Gandhi's theory of nonviolent conflict resolution. Naess has written several books on Gandhi and non-violence. His deep ecology practice included the application of Gandhi's method of nonviolent resistance to defend rivers and waterfalls against the construction of big dams in Norway and a simple, perhaps ascetic, style of living in the cottage Tvergastein in the mountains in the Southern Norway. Naess was deeply influenced by Gandhi's unique method of realizing truth. For Gandhi, realization of truth was attained in a combination of political action and asceticism.

Several elements of Yoga have been incorporated into Deep Ecology through the writings of Arne Naess. This is mainly one of the yogas of *Bhagavadgītā,* Karma Yoga. Naess in his writings calls himself a yogi, an action-yogi, and a follower of Karma Yoga.[2]

Karma Yoga he identifies with the path of Gandhi.[3] Naess is found of quoting a verse from *Bhagavad Gītā* which is about Yoga, the verse 6.29, and which, he argues, expresses in a paradigmatic way the oneness of his personal environmental philosophy, which he calls Ecosophy T.[4] Naess understands this verse of the *Bhagavad Gītā* to say that solidarity with all beings and non-violence depend on widening one's identification and that to see the greater Self means to expand one's identification to include all living beings as one's Self. Verse 6.29 reads:

> *sarva-bhūta-stham atmānaṃ sarva-bhutāni cātmani īkṣate*
> *īkṣate yoga-yuktātmā sarvatra samadarśanaḥ*
> He who is yoked in Yoga and sees the same everywhere, he sees himself in all beings and all beings in himself.

Naess quotes Śaṃkara's commentary to explain the verse.[5] *Bhagavad Gītā* chapter six is about Yoga or disciplined meditation as a means for self-realization. This is the only chapter in the *Bhagavad Gītā* that presents Yoga as somewhat similar to the popular meaning of the word Yoga today, that is, as meditation and bodily postures (in this case sitting with erect body). *Bhakti* is of little importance in this part of chapter 6. This part describes the disciplined person who realizes the self, merges with *brahman*, and sees *brahman* in all beings by means of Yoga. The person possessed of such discipline is said to be immovable, with subdued senses, be the same in cold and heat and unmoved by pleasure and pain. For him, all things and all persons are the same. He abides solitary in a hidden place and performs Yoga fixing his thoughts on a single object, sitting (*āsana*), gazing at the tip of the nose, fixing his thought organ on the self, thinking of nothing at all, and causes all thought to come to rest. He realizes the peace of *nirvāṇa*, the self, becomes one with *brahman*, attains the touch of *brahman* and sees himself in all beings and all beings in himself, and sees the same in all things.[6]

Why is there a frequent use of Sanskrit words such as *ātman*, *jīva*, *advaita*, *svamārga*, *karmamārga*, *karmayogi*, *ahiṃsā*, and a prevalence of words common in Hindu texts in English but reinterpreted (self, self-realization, non-violence), and quoting of Hindu sacred texts (*Bhagavadgītā*, Śaṃkara's commentary on the *Bhagavadgītā*) in the deep ecological writings of Naess? The answer to that question is that this use represents the heritage and influence of Gandhi. Gandhi held that all his political activities were aimed at attaining *mokṣa*. Self-realization is also the key concept in the Ecosophy T of Naess. It is the highest norm of his philosophical system. Naess' deep ecology builds on the philosophy of oneness, exemplified by Naess himself with Śaṃkara's Advaita Vedānta, and with Gandhi. The first principle of Ecosophy T is the same as the first principle in Gandhi's teaching according to Naess: 'Seek complete Self-Realization'.[7] By Self-realization Naess means that when 'the widening and the deepening of the self goes on ad infinitum the selves will realise themselves by realising the same, whatever that is'.[8] But at any level of realization, he argues, the egos remain separate, they do not dissolve like the drops in the ocean. Oneness does not mean absolute nonduality. This is a qualified non-dualism. But he argues that 'the higher the Self-realization attained by anyone, the broader and deeper the identification with others'. The point is to identify with the larger ecosystem and other species in order to realize that nature is not our enemy, as nature often has been portrayed in the West, and something evil that we need human culture to protect us from. The self-realization of other species is part of our self-realization! From the viewpoint of Ecosophy T nature is a realm of play in which all beings try to realize themselves. This is a natural process, and beautifull as well.

Naess also asks who in history have been preeminent in approaching Self-realization. He is critical to asking such a question, but mentions Gandhi as the genius of non-violence as someone who could qualify.[9] Arne Naess builds on a philosophy that is much in agreement with modern reinterpretations of Vedanta. Naess argues that the deep ecology path first goes inward, only to lead out again to everything. This can be compared to Sarvapalli Radhakrishna's commentary on *Bhagavad Gītā* 6.29, the verse often quoted by Naess:

> Though, in the process of attaining the vision of Self, we had to retreat from outward things and separate the Self from the world, when the vision is attained the world is drawn into the Self. On the ethical plane, this means that there should grow a detachment from the world and when it is attained, a return to it, through love, suffering and sacrifice for it. The sense of a separate finite self with its hopes and fears, its likes and dislikes is destroyed.[10]

The Ecosophy of Arne Naess as it is presented in his writings draws on the close connection between non-violence (*ahiṃsā*), the philosophy of oneness (*advaita*) and the goal of self-realization (*mokṣa*) in the religious thought of Gandhi. Self-realization as the goal, oneness of all living beings and non-violent political action is fundamental both to the ecosophy of Arne Neass and to the philosophy of Gandhi.

Sāṃkhya-Yoga and the Disharmony of Nature

Naess has argued that Deep Ecology is a program that can be supported by different ecosophies or 'total views inspired in part by reactions to the ecological crisis'.[11] A number of religions and philosophies can function as ecosophies, but some are perhaps too anthropocentric, and many of them are, traditionally at least, not about the preservation of the natural world. For some religions, it seems, the interest in environmentalism is an attempt to become political relevant and gain support in secular surroundings.

Philosophical monism is at the foundation of the Deep Ecology of Arne Naess, Ecosphy T.[12] How does the Deep Ecology view look from the point of view of the dualist philosophy of Sāṃkhya-Yoga? What would a possible Sāṃkhya-Yoga foundation of Deep Ecology look like? Given the close relationship between Deep Ecology and some Hindu traditions, one would expect that Sāṃkhya-Yoga easily could be constructed as a foundation for Deep Ecology.

Certainly, renunciants consume little and are non-materialistic and in that sense exemplify an eco-friendly style of living. This is an important message from the renunciant traditions of what is important in life. Solitude in nature is highly valued, greed is understood as a mental fault, and accumulation of wealth is looked down upon as distracting from the ultimate goal of life. Wisdom is associated with the mountains, not the cities.

However, with its emphasis on dualism, suffering (*duḥkha*) and *ahiṃsā*, Sāṃkhya-Yoga has a quite different view of nature than the ecosophy of Arne Naess.[13] This ecosophy understands nature as a harmonious interdependent whole with which one should identify and merge to a certain degree. The Deep Ecology of Naess celebrates the happiness of life.[14] Sāṃkhya-Yoga seems to emphasize more the disharmony of nature, and certainly the goal is to withdraw from that disharmony.[15] *Interdependence in Sāṃkhya implies disharmony and difference.*[16] Interdependence in Deep Ecology means oneness and identity. According to Sāṃkhya-Yoga the world is not a harmonious functioning whole, as is emphasized in Deep Ecology, but a disturbed realm. *Mokṣa*, freedom from interdependency, is therefore the goal. The soul, the *puruṣa*, according to Sāṃkhya-Yoga, is different from the interdependent whole of nature. The similes taken from nature in Sāṃkhya texts emphasize difference, not identity.[17]

The idea of a single material principle, *prakṛti,* as constituted by the three *guṇa*s (*sattva*, *rajas* and *tamas*) which is fundamental in Sāṃkhya-Yoga means that due to the presence of *rajas* there is some element of disharmony in every experience. In the material world all is connected to all, and interdependency means mutual disturbance. Harmony according to Sāṃkhya and Yoga would mean peacefulness and non-disturbance. However, Sāṃkhya-Yoga sees nature also as a process of mutual aid and service which can be interpreted in an ecological sense.[18] But nature implies constant injury of living beings (due to the presence of *rajas).* Therefore one should reduce one's activities and ultimately try to become liberated from materiality. Complete withdrawal from materiality is therefore the ultimate act of non-violence or injury, according to Sāṃkhya-Yoga.

Withdrawal from interdependency is an act of non-violence. Jan Gonda has argued that "according to Indian view *hiṃsā* "violence, injury" belongs to nature itself. Ascetic life, ideal life is a conscious negation of the natural or innate tendencies to "violence.""[19] *Ahiṃsā* is not an ecological ethics, but a negation of natural processes. *Ahiṃsā* is a refusal to participate in the interdependency of the natural world. The act of non-injury becomes the first step toward absolute separation from the interdependent transformations of the material world. This follows logically from the dualistic position of Sāṃkhya and Yoga.

In Vācaspati Miśra's commentary on the *Yogasūtra*, even the desirability of attaining heaven at the cost of the life of others is questioned. The criterion for meritorious behavior is that one should not do to others that which is disagreeable to one's own self. 'Others' here means all living beings and not just humans. Every act implies some sort of violence toward others. In nature everything is for the service of everything else. But every act is also accompanied by the bringing of pain to some living beings. You cannot do good to anyone without harming other beings. That is the disharmony of interdependence. *Kaivalya* is a state that is free from 'injury' *(hiṃsā)* because the *puruṣa* has attained absolute separation from *prakṛti*. Knowledge of suffering is at the base of non-injury. Writes a Sāṃkhya-Yoga commentary to *Yogasūtra* 2.30: 'To nourish one's own body with the flesh of another is the chief form of inflicting injury. Besides, seeking one's own comfort inevitably involves causing pain to others.'[20] Thus seeking one's own comfort inevitably hinders the self-realization of others, to use the language of Deep Ecology. Isolation is the goal of Sāṃkhya and Yoga and not the integrity and harmony of the ecosystem.

Non-injury, *ahiṃsā*, is based on this idea of interdependence as a source of suffering and the disharmony of the world. The understanding of nature as food has been central in the Hindu understanding of nature and the close connection of vegetarianism and non-violence has its basis in this perception. [21] That humans have to kill or to have others kill for them to transform nature into food in order to stay alive is symptomatic of the disharmony of interdependence. The doctrine of non-injury can be understood as an attempt to transcend the ecological processes of nature. It is significant that the first part of the eightfold path of Yoga is the practice of non-injury because this signifies the first break with the interdependency of the natural world. The inclination toward non-injury toward all living beings are natural inclinations in everyone, not only the inclination toward injury. The natural inclination toward non-injury toward all living beings is part of the inclination toward 'separation' *(viveka)* and 'isolation' *(kaivalya)* of *puruṣa* from materiality. This is different from Deep Ecology and Ecosophy T. Ecosophy T favors merging with material nature, Sāṃkhya-Yoga favors separation form material nature. Ecosophy T emphasizes identifying oneself with nature, Sāṃkhya-Yoga separation from nature.

The awareness of the impossibility of both being alive and at the same time not hurting other living beings and thereby causing future pain to oneself, is an element to urge the practitioners of Sāṃkhya and Yoga to strive toward 'liberation' *(mokṣa).* Liberation consists of the experience of the absolute difference between the human 'self' or 'soul' *(puruṣa)* and the 'material principle' *(prakṛti).* The goal of Sāṃkhya and Yoga is to have this experience. The human being, by identifying with the natural world, as Deep Ecology declares, misunderstands his or her true identity. There is a transcendent aspect of human beings, as with all other beings, since all beings are a mixture of *puruṣa* and *prakṛti*, consciousness and materiality. The ethics of non-injury toward all beings in Sāṃkhya-Yoga is associated with the knowledge of the painful nature of existence and the idea that the true identity of all beings totally transcends nature. *Ahiṃsā* in Sāṃkhya-Yoga is not an ecological practice, and it is not based on a deep ecological experience. It is based on focusing on suffering, not the ecological harmony of unity, but on *the disharmony of unity.* Sāṃkhya and Yoga do not believe in the possibility to change the natural world in order to make it compatible with human happiness. Rather it is the human being who has to change his or her perception of reality. Nature should be left alone. Here there is some possible agreement between deep ecology and Sāṃkhya-Yoga. Deep ecology favors the establishment of large territories free from human development.[22]

The Idealized Nature of Hindu asceticism

The ability of an ascetic to live alone in wilderness among wild animals and harsh natural conditions is traditionally a sign of sacredness in the Hindu tradition. The peacefulness of the yogin is thought to transform the natural surroundings. One of the supernormal powers that is attained through the ethical perfection of non-injury is peaceful surroundings. Wild animals become peaceful around the ascetic. Places of yogins are often described in the Hindu texts as places of extreme natural beauty. YS 2.35 says: 'As the yogin becomes established in non-injury, all beings coming near him cease to be hostile.' As a consequence of perfection in non-injury, the belief is that the surroundings of the yogin become free from violence.

In the Sanskrit literature, the sacred places of ascetics are described as places of peace in which nature has been tamed and the natural processes such as the seasons have been transcended. In this way does the poet Kālidāsa describe the transformation of place caused by the power of the yogin:

> This is the holy grove of Atri, a means of accomplishing asceticism, in which the wild animals have been tamed without the fear of chastisement, where the trees have been bearing fruits without having put forth flowers...therefore displays the mighty power of the sage.[23]
>
> Even these trees in the middle of the Vedis of the sages who devote themselves to meditation in the Virāsana posture, appear, absorbed in (Yoga) meditation, as it were, on account of the stillness caused by the absence of breeze.[24]

Nature has become transformed by asceticism. This is the idealized nature of Hindu asceticism. This is not nature in an ecological sense but nature improved by the practice of Yoga. It is nature without the food chain and nature without the natural forces, it is a non-natural idealized, what could be called transcendent environment. It is nature tamed by culture. Nature has been tamed by the asceticism of the yogis to become the idealized nature of peace. Also the hermitage of Vālmīki is described by Kālidāsa as such a non-natural idealized environment. It is a place where 'in the evening, the deer were sitting by the side of the altars and the wild animals were in a state of peace and tranquility.'[25] The *ahiṃsā* of the yogin has produced non-violent nature. The natural process of the food chain has been set aside. It could perhaps be called transcendent nature to emphasize that it is nature transformed by religious practice. In some versions of Deep Ecology, it is this idealized view of nature that dominates.

Deep Ecology celebrates nature as the place of self-realization of all species. According to Sāṃkhya-Yoga nature is the place of Self-realization but only those born as humans have the possibility of self-realization. Animals have to be reborn as humans. Humans can be reborn as animals, but that is considered a bad rebirth. Caring too much for animals or the environment can become a hindrance and can even lead to a rebirth as an animal. An example of this is given in the *Sāṃkhyasūtra*. Certainly, one should live alone in nature, but one should not get caught in the love of nature. Book four of the *Sāṃkhyasūtra* contains many advices about how one should live, given in the form of similes.[26] A Sāṃkhya-Yogin should live alone in nature. *Sūtra* 4.9 states that in association with many people there is internal strife because of the manifestation of passion, etc., as in the case of a girl's shell-bracelet (*bahubhir yoge virodho rāgādibhiḥ kumārīśaṅkhavat*) (4.9). The next *sūtra* says that there is disturbance even from the company of two (*dvābhyām api tathaiva* 4.10). The person serious about practicing Sāṃkhya-Yoga should avoid other people and live alone. Only nature offers such opportunities to live alone. In wilderness only can there be freedom from the company other people. So, only nature's company is acceptable for one who wants to concentrate on knowledge leading to liberation. There is an implicit message about nature here. Nature gives rise to passionlessness, freedom from bondage in the world. Nature is the environment in which wisdom is realized. Caring for the environment in a modern environmental sense is irrelevant for traditional Sāṃkhya-Yoga, and can even become a hindrance. This is clearly stated in the *Sāṃkhyasūtra* 4.8. *Sāṃkhyasūtra* 4.8 says that what is not a means towards liberation is not to be thought of (*asādhanānucintanam*) because it leads to bondage (*bandhāya*) as in the example of Bharata (*bharatavat*). The example of Bharata refers to a story about King Bharata, the well known founder of the Bharata Dynasty, and the deer.[27] The royal sage Bharata was performing Yoga and was close to gaining release, but when he watched a female deer die soon after giving birth, he took care of the young deer and thereafter he got so attached to it that he thought of nothing else. His yogic practices had been interrupted and he did not attain release even if he had been almost there! He forgot the reality of the self, fixed his mind on the deer, even at the time of death, and was thus reborn as a deer. As a deer he remembered his previous life and was full of regrets that he had strayed from the path of Yoga.[28] Therefore the Sāṃkhya-Yogins should be free from attachment and should not get involved in things that do not lead to release. This is a clear statement that Sāṃkhya-Yoga is about *apavarga*, *mokṣa*, and not *bhoga* and the environment.

Sāṃkhya and Yoga say that matter can be known in two ways, as 'knowledge for release' *(apavarga)* and as 'knowledge for enjoyment' *(bhoga)*. *Bhoga* is the knowledge of matter for the purpose of using the objects to give 'pleasure' *(sukha)* and 'pain' *(duḥkha)*. This is the normal empirical knowledge of matter and the goal of practices such as art, science, agriculture, industry and technology. This is the utilitarian knowledge of the objects around us, like knowledge of how to transform plants and animals into food or metals into machines, or the aesthetic enjoyment of a beautiful landscape. The active human transformation of matter in arts, agriculture, industry, mining, and so on, is motivated by the wish for enjoyment. Ordinary knowing is the knowing of matter for that purpose. According to Sāṃkhya and Yoga, there is a knowledge of matter that is different from this ordinary, utilitarian way of knowing matter. *Apavarga* is the knowledge of matter for the sake of leaving matter behind, not for the sake of experiencing its pleasures. Deep Ecology knows nature as *bhoga*, Sāṃkhya-Yoga knows nature as *apavarga*. In order to know matter as *apavarga* one must be detached from matter. The realization of the absolute separateness of the material principle and *puruṣa*, which together constitute the *buddhi*, liberates the individual *puruṣa*. The effort aimed at enjoyment of matter and the effort aimed at the leaving of matter give rise to two different kinds of knowledge. One is the ordinary, utilitarian knowledge and the other is the knowledge that leads to *mokṣa*. Most persons, according to Sāṃkhya and Yoga, wish only for the enjoyment of matter, either on earth in this life or in heaven in a future life. Few persons wish for the release from materiality.

In the introduction to *Hinduism and Ecology*, Christopher Chapple refers to an outline of progressive stages that indicate increasing radical commitment to ecological harmony.[29] This scheme exemplifies the difference between Deep Ecology and Sāṃkhya-Yoga.The first promotes development by use of natural resources. This should not really be associated with a commitment to ecological harmony. The second emphasizes utility and sustainable development, i.e., shallow ecology. The third stage is the romantic, it sees ultimate reality manifest itself in nature and respects the divinity of nature. This promotes deep ecology. The fourth stage is asceticism and entails separation from nature through forms of abstinence. It supports the withdrawing from the world and is the way of the renunciants.[30] Sāṃkhya-Yoga belongs to the fourth stage of this scheme.

Consciousness of the disharmony of interdependency can lead to withdrawal from the world. Withdrawal promotes ecological harmony as exemplified in Kalidāsa's description of āśramas in India. It should be noted that the founder of Deep Ecology, Arne Naess, has said that he regrets only one thing in life, that he did not become a Hindu renunciant when he was a young man. This would imply the withdrawal from the human world into wilderness in order to realize separation from society. Naess chose instead Gandhi's Karma Yoga. This longing for wilderness and association of wisdom with wilderness are nevertheless most important elements of Deep Ecology and important Hindu influences on Deep Ecology as well.

A weakness of many varieties of environmentalism is that they promote the same attitude of gaining human control of the environment that is the source of the environmental crisis. The ecological message of Sāṃkhya-Yoga, on the other hand, is that one should withdraw from the world. Leave nature as it is! One should control oneself, not the world!

On the way to *kaivalya* the renunciant passes through stages of idealized nature in which harmony and peacefulness are realized. This is not ecological nature but nature in which the human ethics of *ahiṃsā*, a non-ecological ethics when applied to non-humans, has been projected on all of nature. It is a projection on nature of the ideal of *ahiṃsā* of Sāṃkhya-Yoga. But the harmony and peacefulness of *ahiṃsā* is not nature in an ecological sense. This is nature improved by the practice of Yoga. The last stage of Sāṃkhya-Yoga is *kaivalya*, the final withdrawal from the world. As an ecological practice, this withdrawal is probably closer to the ideals of Deep Ecology than many of the environmentalist efforts of gaining human control of the environment.

Endnotes

[1] Some of the material in this part has previously been published in Knut A. Jacobsen. 'Bhagavadgita, Ecosophy T and Deep Ecology,' in *Beneath the Surface: Critical Essays in the Philosophy of Deep Ecology* (Eric Katz, Andrew Light, and David Rothenberg, eds., Cambridge: The MIT Press, 2000), 231–52.

[2] Arne Naess. *Ecology, Community and Lifestyle: Outline of an Ecosophy* (Cambridge: Cambridge University Press 1989), p. 194.

[3] *Karma-yogi* is here a concept borrowed from Gandhi. Vivekānanda, who was a great influence on how Hinduism was received in the West, held that the Hindu ascetics should perform social service (*sevā*). This ideal he called *Karma Yoga*. Gandhi was a foremost practitioner of this ideal. He made social service a necessity for Self-realization.

[4] Naess calls his own foundation of deep ecology Ecosophy T.

[5] David L. Haberman. *River of Love in an Age of Pollution: The Yamuna River of Northern India* (Berkeley: University of California Press, 2006), criticizes that I, in the article 'Bhagavadgita, Ecosophy T and Deep Ecology,' in *Beneath the Surface: Critical Essays in the Philosophy of Deep Ecology*, (Eric Katz, Andrew Light, and David Rothenberg, eds., Cambridge: The MIT Press, 2000), 231–52, connect Deep Ecology to the interpretation of the *Bhagavad Gītā* by the ascetic traditions and in particular to Śaṃkara, and not the *bhakti* traditions. He notes that *Bhagavad Gītā* was a main text for the Bhāgavatas and the devotional traditions and not only the ascetic traditions. While this is correct, my point in that article was to identify which traditions in Hinduism the deep ecologist Arne Naess was influenced by. I asked the question that since *Bhagavad Gītā* is a devotional text favoring the worship of Kṛṣṇa, why would Naess who do not believe in or has ever shown any interest in Kṛṣṇa or any personal god, quote this text? Whose *Bhagavad Gītā* was Naess interested in? Naess was not influenced by the Bhāgavata

tradition or the devotional tradition but by the interpretation of the philosophers of the Advaita Vedanta (Śaṃkara, Sarvapalli Radhakrishnan) and Mahatma Gandhi. Naess quotes Śaṃkara and Radhakrishnan to explain the *Bhagavad Gītā*. Naess does not refer to Kṛṣṇa nor does he relate to a personal god. Naess' *Bhagavad Gītā* is primarily the *Bhagavad Gītā* of Gandhi and of Śaṃkara. This does not mean that I give primacy to Gandhi's or Śaṃkara's interpretations of the *Bhagavadgītā*. Neither Śaṃkara nor Gandhi were interested in the element of Kṛṣṇa *bhakti* in the *Bhagavadgītā*. There are two different issues involved here: first, which schools of interpretation of the *Bhagavad Gītā* influenced Arne Naess' Deep Ecology (the topic of my article); and second, which schools of interpretation of the *Bhagavad Gītā* have the greatest relevance for contemporary environmentalism. These questions should not be confused.

[6] This disciplined person is similar to the person of stabilized mentality (*sthithaprajña*) described in the Bhagavad Gītā 2.54–72, the verses which Gandhi maintained contained the essence of the Bhagavadgītā, which Gandhi called the *satyagrahi*.

[7] Arne Naess, *Gandhi and Group Conflict*, p. 54.

[8] Arne Naess, *Ecology, Community and Lifestyle*, p. 195.

[9] Arne Naess, *Ecology, Community and Lifestyle*, p. 196.

[10] *The Bhagavadgītā*. With an Introductory Essay, Sanskrit text, English Translation and notes by S. Radhakrishnan, p. 204.

[11] Arne Naess, 'The Deep Ecology Eight Points Revisited', in *Deep Ecology for the Twenty-first Century*, (George Sessions, ed., Boston: Shambhala, 1995), p. 215.

[12] See Knut A. Jacobsen, 'Bhagavadgita, Ecosophy T and Deep Ecology.' In *Beneath the Surface: Critical Essays in the Philosophy of Deep Ecology*, (Eric Katz, Andrew Light, and David Rothenberg, eds., Cambridge: The MIT Press, 2000), 231–52.

[13] Some of the material in this part has previously been published in Knut A. Jacobsen, *Prakṛti in Sāṃkhya-Yoga: Material Principle, Religious Experience, Ethical Implications* (New York: Peter Lang, 1999).

[14] The celebration of playfulness became a main theme in the philosophy of Arne Naess during the last part of his life.

[15] But nature is also an important source of sacredness. The presence of divinities and sacred powers in nature and especially on sacred places and the power of places themselves to grant salvation is also significant for the Indian view of nature. Sacredness of place and disharmony of interdependency are probably the two most significant element in the Indian view of nature.

[16] See Knut A. Jacobsen, *Prakṛti in Sāṃkhya-Yoga: Material Principle, Religious Experience, Ethical Implications* (New York: Peter Lang, 1999).

[17] Knut A. Jacobsen. 'What Similes in Sāṃkhya Do: A Comparison of the Similes in the Sāṃkhya texts of the *Mahābhārata*, the *Sāṃkhyakārikā* and the *Sāṃkhyasūtra*.' *Journal of Indian Philosophy* 34 (2006): 587–605.

[18] Jacobsen, *Prakṛti in Sāṃkhya-Yoga.*

[19] Jan Gonda, *Four Studies in the Language of the Veda.* 'S-Gravenhage: Mouton, 1959, p. 97.

[20] Hariharānanda Āraṇya, *Yoga Philosophy of Patañjali*, p. 209.

[21] India is the only country in the world in which vegetarianism is practiced by more than a small minority (probably more than 30% of the population are vegetarians). Vegetarianism is connected to the idea of non-violence (*ahiṃsā*) and the view of nature as food. Non-violence in India was in origin a reaction to the killing of animals, that is, about the relationship to nature, and not a reaction to war.

[22] Arne Naess, *Ecology, Community and Lifestyle*, p. 212.

[23] *The Raghuvamśa of Kālidāsa with the commentary of Mallinātha*, ed. and trans. Gopal Raghunath Nandargikar, 4th ed. (Delhi: Motilal Banarsidass, 1971), 13.50, p. 416.
[24] Ibid., 13.52, pp. 416–17.
[25] Ibid., 14.19, p. 457.
[26] Jacobsen, 'What Similes in Sāṃkhya Do'.
[27] The story is told in *Bhāgavata Purāṇa* Book 5, chapter 8.
[28] *Bhāgavata Purāṇa* 5.8.27–29.
[29] 'Introduction' in *Hinduism and* Ecology*: The Intersection of Earth, Sky and Water* (Christhopher K. Chapple and Mary Evelyn Tucker eds., Cambridge: Harvard University Press, 2000), p. xliii.
[30] 'Introduction' in *Hinduism and Ecology*, p. xliv.

Towards a Theory of Tantra-Ecology

Jeffrey S. Lidke

Eco-Scholarship on Tantra

Ecosophy is a broad-based movement that utilizes the findings of ecology as the foundation for philosophical reflection and spiritual practice rooted in environmental activism. A predominant characteristic of ecosophy is the belief that personal freedom is inseparable from the well-being of the natural world. In an important comparative essay titled "Bhagavad Gita, Ecosophy T, and Deep Ecology," Knut Jacobsen carefully delineates the distinctions between self-realization in Indian monastic traditions and the aspirations of ecosophists:

> In environmentalism, preservation of *saṃsāra* has become the ultimate goal, and not liberation from it, because the preservation of *saṃsāra* is seen as identical with realization of oneself. While self-realization in the monastic tradition meant ultimately to be free from the biological life and death cycle, self-realization in environmentalism means the flourishing of biological life. Environmentalism values the biological world of birth and death (*saṃsāra*) as ultimate reality, not a changeless substratum. Self-realization is nothing else than *saṃsāra*, and the realization of *saṃsāra*, i.e., to identify oneself fully with the natural processes, is *mokṣa*. The context is not liberation of the self from *saṃsāra* but liberation of the natural world from the suffering caused by human beings ignorant of the true identity of the self. The unity of all being does not mean that all beings share

> the same self, as in Advaita Vedānta, nor the oneness of humanity, as often in political interpretations, but organic wholeness, interdependence, the experience of sharing the joy and suffering of all living beings, looking at their self-realization as one's own. The definition chosen by Ecosophy T for living beings is the Hindu definition, namely beings capable of self-realization, i.e., those sharing or possessing an *ātman* or *puruṣa*.[1]

As Jacobsen himself makes clear, ecosophical traditions ought not to be categorized as the 'same' as any particular Indian *darśana*. Ecosophy is rooted in the complex and context-specific political agendas of 20th century activists, scientists, and philosophers—agendas that bear the mark of a genealogical tree whose intellectual roots extend into soils far afield from the land in which yogis reflected on how to achieve liberation. Nonetheless, the broad based similarities between ecosophical thought and some Indian philosophical reflection has been so apparent to certain ecosophists that they have enthusiastically drawn from Indian scriptural sources—primarily the Bhagavad Gītā and the writings of Gandhi—in order to reinforce the foundations of their own discourse and agendas.[2]

Jacobsen is not the only scholar to explore the possibilities for a two-way exchange between ecosophy and Indian thought. In his important work on Yoga, *Integrity of the Yoga-Darśana*, Whicher puts forth an insightful 'ecosophical interpretation' of classic Yoga in carefully arguing that *samādhi* does not culminate with the recognition of a 'disunion' of nature (*prakṛti*) and consciousness (*puruṣa*), but rather in their discriminative integration. The stilling of the thoughtwaves of the mind in higher states of meditative absorption does not lead, in the end, to a rejection of the world; rather, Whicher argued, it enables the yogin to attain a state of equipoise and insight that enables him to master and playfully engage the 'modifications of nature,' seeing them as a continuation on a spectrum of consciousness that extends from 'matter' to 'spirit'. In this way, Whicher argued, Yoga highest result is in an affirmation of the body, the natural world and the yogin's identification with both.

Whicher's critique of predominant scholarly interpretations of Yoga as anti-environmental, dualistic discourse is in alignment with the analysis of eco-feminist scholar, Vandana Shiva:

> Contemporary Western views of nature are fraught with the dichotomy of duality between man and woman, and person and nature. In Indian cosmology, by contrast, person and nature (Purusha-Prakriti) are a duality in unity. They are inseparable complements of one another in nature, in woman, in man. Every form of creation bears the sign of this dialectical unity, of diversity within a unifying principle. . . . Since, ontologically, there is no dualism between man and nature and because nature as Prakriti sustains life, nature has been treated as integral and inviolable. Prakriti, far from being an esoteric abstraction, is an every day concept which organizes daily life. . . . As an embodiment and manifestation of the feminine principle it is characterized by (a) creativity, activity, productivity; (b) diversity in form and aspect; (c) connectedness and inter-relationship of all beings, including man; (d) continuity between the human and natural; and (e) sanctity of life in nature.[3]

Shiva's analysis reflects a common theme in eco-feminist and eco-theological thought: that the natural world is the dynamic body of the Divine Feminine, a body that is creative, diverse, interwoven, and sanctified. Shiva articulates a 'hermeneutics of interconnectedness' that links 'nature' to 'man' by positing a 'duality-in-unity' relationship of humanity predicated on the notion of a 'connectedness and inter-relationship of all beings.'

In the brief reflections to follow, I wish to think through the possibilities for 'thinking with' ecosophical thought as a creative and comparative hermeneutical exercise aimed at refining and expanding our understanding of one particular form of yogic thought and practice: Śākta Tantra. In so doing, I will not be the first scholar to address the intriguing similarities between Śākta Tantra and ecological writings and practice. This distinction belongs to Rita Dasgupta Sherma who first addressed the topic in 1998 in an important essay "Sacred Immanence: Reflections of Ecofeminism in Hindu Tantra." In this work, Sherma demonstrates, like Jacobsen, that much Indian spirituality does not share the ecologists' goal of preserving nature or even finding knowledge through some kind of reflection on the 'nature of nature.' For many Hindus, Sherma reminds us, nature has been identified as a source of bondage. However, Tantra, she argues, represents an alternative hermeneutical trajectory within Indic traditions that affirms the embodied world as the field of liberation. In this regard, she articulates seven 'affinities' between Tantra and ecological thought: (1) celebration of all aspects of life (2) elevation to ultimacy of a feminine principle linked to materiality; (3) possibility for liberation of female gender from constraints of 'fertility and nurturance alone'; (4) affirmation of phenomena as Goddess; (5) articulation of a discourse of empowerment for the marginalized; (6) veneration of the body and its sensations; (7) absence of a spirit/matter dichotomy.[4]

Sherma goes on to define Tantra as a 'theology of identification' that is not only "helpful for the cultivation of an earth-centered spirituality" but also a source of "inspiration" for "personal spiritual empowerment."[5] However, she is careful to point that Tantra can easily be misused for power-centered agendas that have little to do with affirmation of the environment.[6] Much of Tantra, she observes has been used for various 'nefarious' purposes linked to sorcery and the affairs of state. Even still, she concludes, a "reconstruction" of Śākta Tantra "can become a channel through which Hindu nondualism can inspire a viable philosophy on which to base a transformed vision of the earth."[7]

Sherma's scholarship embodies an aesthetic beauty and pragmatic concern that I very much appreciate. However, while I am in agreement with much of Sherma's argument, my own interest herein is not to further inspire such a "transformed vision." Rather, my aim is much more circumscribed. In the remaining pages, I attempt to reflect back on Tantric thought and practice utilizing the modern insights of ecosophy. I seek not to suggest that Tantra is the same as ecosophy. But rather, in the spirit of J.Z. Smith I seek to observe how their revealed differences can be illuminating.

Turning to the Texts

Clearly, Tāntrikas were not seeking to save rivers and trees and regulate factories. The post-industrial devastation of our natural resources was not a concern in the world of the founders of Tantra. Moreover, many Tāntrikas—as White, Davidson, Urban, Dyczkowski and other scholars of Tantra have clearly documented—were seeking an empowerment that was firmly rooted in social and political constructions of the 'nation' (*maṇḍala).* In these contexts, power (*śakti*) was interpreted and wielded in relationship to those strategies of state that had less to do with spiritual fulfillment and ecological well-being and more to do with the acquisition of lands and the control of peoples.

However, as White demonstrates in *Kiss of the Yogini*, by the medieval period Tantra was defining not only the 'center' (i.e., the ideology and statecraft of the political elite) but also the 'periphery' (the ideology, counter politics, and spirituality of the populace). In this way, Tantra came 'interweave' the interests of politicians and generals with the high priests, philosophers, artists, poets, mystics and 'common folk' as a total system of knowledge linking meditation, health, worship, the arts, and statecraft through a discourse and logic that made the universe meaningful by highlighting its multileveled interconnections (*bandha*s).

Herein, I am limiting my concerns with that circle within the Tantric *maṇḍala* which contained the aspirations of an elite section of the population privileged to receive initiation into, practice, and write on Tantric yogic practice or *sādhana*. It is within this circumscribed field of the greater Tantric system, that one can identify an 'ecological logic' rooted in the supreme goal (*paramārtha*) of experiencing a liberating empowerment through the embodied world. Akin to Whicher's interpretation of Yoga as 'integration', I suggest that we interpret this high Tantric aspiration, not as a disembodied experience of 'pure consciousness' but as the concrete and literal embodying of the ecosphere.

In the highly coded environs of Tantric practice the final aim is the realization that the body of the *sādhaka* and the body of divinity are united in a holographic universe[8] whose constituent parts contain within themselves the whole, "this all" (*sarvaṃ idaṃ*). The *Śiva Saṃhitā*, a Nāth Siddha guide to Haṭha Yoga (ca. Tenth century), describes the body of the *yogin* as the seat of the entire universe.

> In your body is Mount Meru, encircled by the seven continents; the rivers are there too, the seas, the mountains, the plains, and the gods of the fields.

> Prophets are to be seen in it, monks, places of pilgrimage and the deities presiding over them. The stars are there, and the planets, and the sun together with the moon; there too are the two cosmic forces: that which destroys, that which creates; and all the elements: ether, air and fire, water and earth. Yes, in your body are all things that exist in the three worlds, all performing their prescribed functions around Mount Meru; he alone who knows this is held to be a true yogi.[9]

In Śākta Tantra the Goddess is celebrated as manifesting simultaneously on the macrocomic plane as the universe and on the microcosmic plane in the human physiology. In Tantric traditions this twofold manifestation is at times described as a "double concealment" in which divine consciousness conceals its true nature. Sanjukta Gutpa remarks:

> Tantric philosophy says that ultimately the unconscious bits of the universe, like stones, are also God and hence consciousness that has decided to conceal itself (*ātma-saṃkoca*). Here we come to the double concealment which God decides on: firstly, He conceals the fact that His true form is identical with the individual soul; and secondly, he conceals His true nature as consciousness to manifest Himself as unconscious phenomena.[10]

The Absolute's contraction as the universe is understood in this context as the outward projection of its inner nature.[11] In this non-dual perspective, the universe is not a limitation of the Godhead. Rather, it is the pristine reflection of its infinite creative powers (*ananta-kalā-śakti*). The Godhead becomes the universe and all beings in it, enfolding[12] itself into an infinitely varied cosmic dance. However, once manifested as all living beings, the Godhead in each case conceals its true nature (*svarūpa-saṃkocana*). Tantric ritual and yogic practices provide the tools for the *sādhaka* to awaken to his or her true nature as that supreme consciousness-power which is the source and goal of all creation.

The key to achieving this realization is initiation into a Tantric lineage of perfected ones (*siddha-sampradāya*) stemming directly from the mouth of the Godhead (*divya-mukha*) and capable of revealing the technologies of self-perfection. Initiation includes training in the specialized ritual and yogic procedures that produce transformations in consciousness as a result of the manipulations of the fluids of the physical body and the energies of the subtle body. Across sectarian divisions, Tantric systems of *sādhana* share certain common features. In each case, the aim is to reverse the process of cosmogenesis and return the Godhead's projected manifestations back to their unmanifest source. During *sādhana* the practitioner encodes in his or her microcosmic form the various parts of the Godhead's macrocosmic form: divinities (*devatā*s), phones (*mātṛkā*s), graphemes (*kāra*s), elementary principles (*tattva*s), worlds (*loka*s), and I-cognizers (*pramātṛ*s).[13] In this way, the *sādhaka* reproduces the process of cosmogenesis within his or her own psychophysiology. He or she then reverses this process by harnessing the regressive power of the *visarga-śakti*[14] and awakening the *kuṇḍalinī-śakti* seated at the base of the subtle physiology. Once awakened, the *kuṇḍalinī-śakti* ascends through the central channel, its ascent representing the dissolution of the universe in which all manifest forms are absorbed back into their unmanifest source in Paramaśiva at the crown of the head.

The mechanics of the *sādhaka*'s reversal of the cosmogonic process and return to the Godhead function according to an internal-external dialectic in which modalities of external worship (*bahir-yāga*) are mirrored by internalized visualizations and yogic practices (*antar-yāga*).[15] The template that mediates this dialectic is the *yantra*, the mesocosmic device that is imparted by the *guru* at the time of initiation, *dīkṣā*.[16] The *yantra* is the geometric embodiment of the divine that functions simultaneously as the image of the divinity, the image of the universe, and the "image of man."[17] The Tantric *sādhaka* employs this mesocosmic device in both external ritual worship (*pūjā*) and internal meditative practice as a means of tracing the unfoldment of the cosmogonic process (*sṛṣṭi-krama*) from the *bindu* in the center to the outer circuits of the *yantra*'s periphery and, conversely, as a means of reversing the cosmogenesis by tracing the process of dissolution (*laya-krama*) starting from the periphery and moving inward to the center, the *bindu*. The adept's external ritual actions are mirrored by an internal movement of consciousness in which he or she moves from an extrovertive state of multiplicity represented by the *yantra*'s outer circuits to an introvertive state of undifferentiated unified awareness represented by the *bindu* in the center. In the advanced stages of *sādhana*, this movement in consciousness is accompanied by the movement of the *kuṇḍalinī–śakti* from the *mūlādhāra-cakra* at the base of the spine to the *sahasrāra-cakra* at the crown of the head, which is identified with the *bindu*. Once the *kuṇḍalinī* reaches its final destination and becomes permanently established in the *sahasrāra-cakra*, the practitioner becomes a *siddha*, enters the "non-way" (*anupāya*), and transcends the need for any further form of practice.[18]

In the specialized *sādhanas* based on the *Nityāṣoḍaśikārṇava*[19] the *sādhaka* transforms his body into the cosmogonic blueprint by meditatively constructing the Śrī Yantra within her psychosomatic landscape. This process identifies the *yonī* as the inner triangle of the *yantra*, the womb of consciousness from which all creation arises and is paradoxically situated at the base of the abdomen of the human being. And the functions, qualities, and design of this organ are to be seen as the supreme symbol of divine consciousness. Its fires, excretions, and juices are the alchemical properties and extractions necessary for the transformation of a microcosmic consciousness into the macrocosmic being that is Tripurasundarī. This potential for a transformation rooted in human sexuality is the supreme secret (*mahārahasya*) at the heart of high Tantric practice. It is this secret that became institutionalized as Nepal's Kumārī, the virgin goddess whose lower mouth (*adhavaktra*) was the gateway through which kings accessed total power. And it is this secret, encapsulating the paradox of being and replete with liberating power, that Tāntrikas unravel for the purpose of transforming themselves into the absolute. This secret is the mystery of God's capacity to conceal himself from himself, to be one while perpetually manifesting as more-than-one, to project limitation into transcendence and cloak omnipotence with weakness. This discourse of non-duality—the product of at least two thousand years of subcontinental theologizing—creates, through a baroque body technology, the possibility for embodying

śri yantra

(lit., 'swallowing') the ecosphere (*brahmanda*). Tantric *sādhana* is a body language, replete with signs whose referent is internalized experience, a realm of cognitive awareness at levels of speech that actualizes the connections (*bandhas*) with the constituent elements of the embodied universe. These connections are in turn channelized as the *sādhaka*'s own I-awareness, identified as the non-discursive field which is the foundation for those mantric cognitions that are the natural world. This I-awareness is the goddess, the divine mother, the foundational *prakṛti*, from whose interconnected and all-encompassing womb (*yonī*), expressed as a Śrī Yantra, the ecosphere arises.

Within this I-awareness lies the seeds of omnipotence in the form of the phonemes of the Sanskrit alphabet, encoded into all creation as a result of consciousness "vomitting out" (*vamiti*) its inner nature. These phonemes, the *mātṛkā*, are the atomic-essence

Tripurasundarī

of Goddess. For this reason, she is invoked as Mātṛsadbhāva, She Whose True Being is the Phonemes. For Nepalese Śrī Vidyā Tāntrikas the key to success in their practice is the realization that this Goddess is the consciousness-power, the dynamic energy, at the heart of all language and that that language is what gives rise to the natural world. In the pursuit of this literal embodiment of 'nature', the *sādhaka* trains himself to perceive all forms of knowledge and power as arising from the very syllables he visualizes as instilled within the *yantra* that is his body. In this way, the *sādhaka* identifies his practice as the integration of all that exists within the natural world. It is this pre-10th century practice, I would argue, that leads to the discourse of the 'natural way' (*sahaja-yāna*). For it is through the internal tasting of the divine nectar (*divyāmṛta*) emitted spontaneously from within the *sādhaka*'s own eco-spherized body, that he collapses all constructed notions (*kalpita-vikalpa*) into a natural, integrated vision of all reality as the playful expression of his own self-nature. "Paramaśiva," writes Amṛtavāgbhava, a 20th century initiate of both Śrī Vidyā and Trika Kaula, "having eternally risen as a wondrous, divine authority, excels all. Through His own exuberant play, He manifests His own Self in the form of the universe."[20] This contemporary *śloka* poetically and poignantly expresses a contemporary Tāntrika's vision of reintegration, a vision woven upon a loom whose warp and weft interweave the constituent elements of the natural world within the tapestry of a body thereby transformed into the *maṇḍala* that is the ecosphere.

In this way, Tantra articulated an understanding that the body is intertwined with a variety of forces that are simultaneously 'outside' in the natural world and 'inside' the individual psychosomatic complex. These forces, personified as a host of deities, are the constituents of a godhead that oscillates between transcendence (*viśvottirna*) and immanence (*viśvamaya*), between being unmanifest and manifest. The template for this oscillating process is the *maṇḍala*, the theology of which is best compared to the hologram: a blueprint comprising self-replications as its atomic structure. In Hindu contexts throughout the globe, the smallest unit (*anu*) and the absolute (Brahman) are fundamentally one.

Thinking according to this particular yogic 'eco-logic', Tāntrikas construct, play within, and liberate themselves by means of a discourse predicated on the notion that the world around them is an outward projection of their own multi-leveled body. The key to their self-liberating strategies was the identification of 'being' with 'sound.' For the Tāntrika, all that exists is sound. Every thing is *mātṛka,* divine, creative sound. The universe arises from the *mātṛka* and is imbued with its energy. The word is within and without. Within the body is the word. In the world is the word. The elements are within the body and within the elements is the word. To effect change in the surrounding world one utilizes the energies of one's own body which is an exact replica thereof. The Tantras abound with examples of this kind of thinking.

The Sarvasiddhi-Stavah, or opening 12 stanzas of the *Nityāṣoḍaśikārṇava*, is paradigmatic. The first *sūtra* equates the Goddess with the cosmos. The next *sūtra* with the letters and then with the body. The text outlines a technology by which the practitioner encodes his body with the natural world. In standard *nyāsa* practice the *sādhaka* raises the *kuṇḍalinī* by returning each of the five elements to their respective higher or more subtle point of energy: earth into water, water into fire, fire into air, air into ether, ether in the void, the void into the absolute. These elements reside within power wheels within the body. They are not, for Tantrikas, simply imaginary. These are the microcosmic correlates of the outer world. The two are inseparably linked.

The result of the encoding of this technology onto and within the body is that the *sādhaka* comes to witness a condition of internal pervasion (*samāveśa*) in which he is encoded as the natural universe.

He is the stars, the constellation, the moon, the five elements, the seas, and the ocean. This is not a metaphorical condition. He is not as vast as or like the sky. He is, in his ritualized fullness, the sky itself. The channels within do not flow like the rivers of his geological surroundings. They are those rivers.

The *Vāgmatisahasranāma Stotram* from the Himavat Khaṇḍa[21] is a hymn praising Nepal's Bagmati River, which is a tributary of the Ganges. The hymn poetically articulates an Indic deep ecological thinking. The river is praised as flowing from the middle of the *maṇipura-cakra* (VVS 134a), as being the Mistress of Yoga, whose essence is Yoga (VVS 142a), as the central *yoginī* of the *svadhiṣṭhāna-cakra* (147a), and as the foundation of the six *cakra*s (159a-b). These verses take us into a world in which geological rivers flow from mountain tops directly into the bodies of yogis and back. Does the world contain the yogi, or the yogi the world? Architects of an Escherian discourse, the authors of Tantric texts challenged their contemporaries to deeply contemplate these questions. Whether the inside is really the outside or not, the key point is that both are part of one greater web of being in which center and periphery inner and outer, above and below are not only deemed relevant to the perspective of the Gazer, but, perhaps more importantly, identified as equally 'real', interwoven expressions of a singular universe that is both 'material' and 'spiritual.'

Conclusion: Envisioning an Eco-Tantric World

There is potential within the ideologies and practices of Hindu traditions for the modern construction of a function eco-philosophical world-view grounded in yet transcendent to its social and cultural roots. The cult-specific dimensions of Tantra require a degree of initiatory secrecy that is beyond our concerns here. In this closing section, I seek to tease out the parameters and lineaments of Tantra-Ecology as my own rudimentary reconstruction of ecosophical thought within and through Tantric traditions. Towards this end, I seek to suggest three principles as the foundations of Tantra-Ecology.

Principle 1: Gratitude for Being in the World

The important scholarship of Jacobsen, Whicher, Shiva, Nelson, Sherma, and Chapple has skillfully assessed the depths and potential for ecologically-oriented thought and practice in Indic traditions. As these scholars have all pointed out, while Indic traditions do contain an impulse towards liberation via transcendence of the world, there is, as discussed above, an arguably deeper impulse towards liberation within and through the embodied forms of the world. This latter, stronger, impulse is clearly present in the majority of Tantric ideologies and practice. It is an impulse grounded in the understanding that the natural world is embodied

Vagmati River

divinity. If misperceived, this divinity, Goddess Earth, may be a cause of bondage and suffering *(māyā)*; however, it is not Earth in her multiple forms that is the cause of delusion. Rather, the cause is the misperception, the fundamental cognitive error that fails to identify our deep, inevitable and necessary connection to the embodied world in which we live. It is this cognitive error, the Tantras tell us, that leads to our sense of being trapped with and bound by our world-experience. The solution is not to escape from the world. After all, is there really another place to go? Even if so, can we be certain that that other realm is a better place to be? Tantra informs us that while there are multiple realms all of those realms are contained within the embodied cognitive structures of the human experience. Moreover, and perhaps most importantly, many tantric traditions take the stance that birth as a human being puts the soul (*jīva*) in the best possible circumstance for realizing its purpose and full potential. Dilgo Khyentse and other Tibetan lamas refer to a human birth as 'precious', as a gift of infinite value. The first principle to be drawn from the study of Tantric traditions in framing a Tantra-Ecology is precisely this sense of preciousness. We are called by Tantra to see our human condition as a blessing to be grateful for. While many of us may not be drawn to engage in Tantric practices, we can perhaps learn to deepen our sense of gratitude for being in these human bodies here an now in the world. And out of this sense of gratitude for having the bodies that we do, we can then proceed a greater awareness and appreciation of our dependence on and interconnection with the diverse web of life that surrounds and sustains us.

Principle 2: Cultivating a Maṇḍalic Vision of Interconnectedness

Beginning with a basic sense of appreciation to our human condition as precious, Tantra-Ecology then encourages individuals to acquire a deepened awareness of the complex web of relations that sustain each of us individually and as a species. In Tantric traditions, the 'awareness of interconnectedness' is cultivated via the technologies associated with the *maṇḍala. Maṇḍala*s—regardless of their cult-specific natures—highlight a pan-Tantric vision of the cosmos as a balanced, harmonious web of interconnections. This 'harmonious vision' functions as a template for city construction, architectural design, and, perhaps most importantly, meditative practice. These archetypal patterns are utilized by Tantric teachers as a means for training their students to bring the energies, fluids, and 'divinities' of their bodies into a state of alignment, not just internally, but with the energies, fluids, and divinities of the external world.

Maṇḍala technologies in Tantric traditions are grounded in the understanding that there is an innate beauty and symmetry in the world and that our individual and collective happiness is not possible in the absence of an understanding of that beauty and symmetry. While Tantric *maṇḍala*s are highly coded symbols requiring initiation, we can nonetheless take from them a basic point: effective ecological practice requires an interdependent vision of ourselves in relation to the world. If we view ourselves as isolated from the world we are less inclined to act in ways that preserve the web of relations that in fact sustain us. Beginning with an appreciation of the preciousness of human birth, we then move to a vision of the interdependence of our 'being' to the web of beings—human, plant, animal, atmospheric, 'cosmic'—that constitutes the fuller network of Life. This vision—embodied in the *maṇḍala*—fulfills itself in the third principle: practice.

Principle 3: The Practice of Cultivating a Sense of Connection to Local Landscapes

In Tantra-Ecology, the first principles of appreciation of our human birth and maṇḍalic vision of interconnectedness fulfill themselves in the third and final principle: the commitment to practices that cultivate our connection to local landscapes. Such practices deepen appreciation and expands vision. Together with the first two principles it forms a triumvirate feedback loop that mimics the circular flow of consciousness and energy mapped in *maṇḍala*s which in turn resembles the circular patterns characteristic of Nature Herself. The fundamentals of this practice are grounded in the conscious effort to acquire a sense of harmony and alignment with one's local landscapes.

Tantric texts instruct its adherents to engage in spiritual practice at rivers, lakes, trees, mountains, caves and other natural sites as they are considered to be places endowed with living power or *śakti*. Tantra-Ecology asserts that the intention to be in nature is a practice that fosters one's appreciation for being alive and deepens one's vision of interconnectedness.

The majesty and beauty of our natural world is inescapable if we simply chose to be in the presence of its extraordinary manifestations. The 'world out there' according to Tantra is fundamentally the same as the 'world in us'. The same five elements that compose its body—earth, fire, water, wind, ether—constitute our own. Making the intention to encounter the sacred in our local landscape—walking in the woods, sitting silently beneath a nearby stream or tree, for example—inevitably deepens one's awareness of the ways by which our own internal processes and cycles mirror and depend on the of the natural world around us. This deepening of awareness is beneficial not only for individual 'spiritual growth' but leads to a way of 'being in the world' that is supportive of the patterns of balance and harmony that reflect the maṇḍalic design of things as they were intended to be.

Endnotes

[1] Knut Jacobsen. "Bhagavad Gita, Ecosophy T and Deep Ecology." In *Inquiry: An Interdisciplinary Journal of Philsophy and the Social Sciences* Vol 39 (June 1996), pp. 233–234.

[2] Jacobsen, "Bhagavad Gita," pp. 230–234. Cf. David Kinsley, *Ecology and Religion: Ecology Spirutality in Cross-Cultural Perspective* (New Jersey: Prentice Hall), pp. 184–192).

[3] Vandana Shiva. *Staying Alive: Women, Ecology, and Development* (London: Zed Books, 1989). Quoted in Roger S. Gottlieb, ed. *This Sacred Earth: Religion, Nature, and Environment* (New York: Routledge, 1996), pp. 383.

[4] Sherma 1998: 123–124.

[5] Ibid., 124.

[6] Ibid., 126.

[7] Ibid., 126.

[8] See Ken Wilber's discussion in *The Holographic Paradigm and other Paradoxes* (Boulder: Shambhala, 1982). Cf. Paul Muller-Ortega, "Tantric Meditation: Vocalic Beginnings," in *Ritual and Speculation in Early Tantrism: Studies in Honor of André Padoux* (Teun Goudriaan, ed., Albany, SUNY Press: 1992), pp. 227–229.

[9] *"Śiva-Saṃhita* 2.1–2.5. Quoted by Jean Varenne in his *Yoga and the Hindu Tradition* (Chicago: University of Chicago Press, 1973), p. 155.

[10] Sanjukta Gupta, "The Maṇḍala as an Image of Man," in Richard Gombrich, ed., *Indian Ritual and its Exegesis* (Delhi: Oxford University Press, 1988), p. 32–41.

[11] Utpaladeva, *Īśvara-pratyabhijñā-kārikā* 4.1. Translated with commentary by B. N. Pandit (Delhi: Motilal Banarsidass Publishers Pvt. Ltd., 2004): pp. 191–192.

[12] I adopt this terminology from Paul Muller-Ortega's discussion in "Tantric Meditation."

[13] Gupta et. al., *Hindu Tantrism,* p. 184-185. Cf. Pandit, S*pecific,* p. 39–52.
[14] Muller-Ortega, "Power," p. 44.
[15] *Nityāṣoḍaśikārṇava* 5.6: *Dhyātvetyādi. Bāhyārcanāntarārcaneti dhyāne yoge 'nāhata-prasphurat-pūjācakrarājācakrarājasannihitaṃ paradevatāṃyathāvadārādhya prāguktaphalāptaye japet.* For a detailed discussion of this internal/external dialectic see Gavin Flood's discussion in his *Body and Cosmology in Kashmir Śavism* (Lewiston: The Edwin Mellen Press, 1993). Cf., Vrajavallabha Dviveda, "Having Becomes a God, He Should Sacrifice to the Gods," in *Ritual and Speculation in Early Tantrism*, p. 127.
[16] See Alexis Sanderson's "Maṇḍala and Āgamic Identity in the Trika of Kashmir," in André Padoux, ed., *Mantras et diagrammes rituels dans l'hindouisme* (Paris: Editions du CNRS, 1986): 169–207. Cf. Dirk Jan Hoens, "Transmission and Fundamental Constituents of the Practice" in *Hindu Tantrism*, p. 808–83.
[17] This is Sanjukta Gupta's terminology. See her "The Maṇḍala as an Image of Man."
[18] B. N. Pandit, "Yoga in the Trika System," in *Specific Principles of Kashmir Śaivism* (New Delhi: Munshiram Manoharlal, 1997): 99. See also: Deba Brata Sensharma's overview of *sādhana* practice in his *The Philosophy of Sādhanā (*New York: SUNY, 1990). Here again, we find parallels with cosmogenesis: depending on one's perspective God's appearance as the universe is either a hierarchical and linear unfolding or an instantaneous self-manifestation. See B. N. Pandit's discussion in his, "Theistic Absolutism and Spiritual Realism," in *Specific Principles of Kashmir Śaivism*, p. 15–28.
[19] *Nityāṣoḍaśikārṇava Tantra.* With the commentaries *Ṛjuvimarśinī* by Śivānanda and Artharatnāvalā by Vidyānada. Edited by Vrajavalabha Diveda. Yogatantragrantham A1A no. 1, Varanasi: 1968.

---

[20] Śaivācārya Amṛtavāgbhava, *Ātmavilāsa*. Translated by B. N. Pandit in his *Specific Principles*, p. 149. Amṛtavāgbhava was a widely respected *guru* of the Trika Śrī Vidyā Kaula schools and the teacher of B. N. Pandit.

[21] *Himavatkhanda*. In *Himalayako Pauranic Itihasa*. (Kashi: Gorakha Tilla, 1956), Chapter 181: pp. 337–342.

Green Yoga: Contemporary Activism and Ancient Practices: A Model for Eight Paths of Green Yoga

Laura Cornell

A new movement is growing in the United States that hopes to bring ecological awareness to both Yoga practice and Yoga theory. The concept of a "Green Studio" for the practice of *Haṭha Yoga* has become commonplace within just the past few years, Yoga mat manufacturers now offer alternatives to the highly toxic PVC (polyvinyl chloride) mat, popular Yoga magazines regularly feature articles on environmental themes, and senior teachers frequently refer to connection to the planet as central to the practice of Yoga. All of these changes reflect the broad awareness of environmental necessity growing in our culture, and an increasing desire for Yoga to be relevant to healing and protecting the earth.

The dissertation research on ecology and Yoga that I carried out from 2003–2005 at the California Institute of Integral Studies helped catalyze what was already a strong potential for greater ecological awareness in the United States Yoga movement. This led to the formation of a non-profit organization, The Green Yoga Association, which played a powerful role both in gathering those with a concern for ecology and Yoga, and in making the connections visible in the wider community. This article will begin by sharing the story of the birthing of the Green Yoga Association, describing several of the initial projects we undertook. The remainder of the chapter will consider the actual practice of Green Yoga, proposing a model that connects the contemporary practice of an ecologically-attuned Yoga with Yoga's ancient paths.

This dissertation research topic was guided by a vision to help bring the ecological roots of Yoga into the present. Even though ecological awareness is inherent to Yoga, it was minimized in the translation of Yoga to the West. In emphasizing the scientific over the mystical aspects of Yoga for modernity, much of Yoga's ecological embeddedness was left out, including stories and practices that involved animals, plants, the elements, the seasons, and even the sun and moon. The drive for outward mastery of the physical body also contributed to a de-emphasis of Yoga's spiritual, values-oriented side, including its environmental relevance. Given today's environmental imperative, the process of re-discovering and re-interpreting Yoga's ecological potential in a way that is relevant for today is a task of timely significance.

To explore how the ecological aspects of Yoga could be enhanced in both teaching and personal practice, Six Kripalu Yoga[1] teachers gathered for a collaborative research project. Two extended retreats, one in 2003 and another in 2004 served as a vehicle to share specific practices and ideas for activating Yoga's environmental potential. In the year between the retreats, personal experiments in teaching and practice were conducted by the participants.

In addition to experiments in eco-Yoga, the six teachers chose to engage the broader community in a dialogue over Yoga's ecological relevance. A short statement was circulated to the nation's top Yoga teachers and scholars, asking for their comments and feedback. This resulted in the completion of a negotiated document, "Yoga is Ecological," which later became known as the "Green Yoga Values Statement." The statement was circulated widely to Yoga teachers and scholars for their endorsement, distributed at Yoga and environmental conferences, and later printed in several Yoga magazines and linked through Yoga studio websites to the Green Yoga Association website. The statement reads as follows:

> The health of our bodies depends on clean air, clean water, and clean food. Yoga is grounded in an understanding of this interconnection. Historically,

> Yoga developed in the context of a close relationship with the earth and cosmos and a profound reverence for animals, plants, soil, water, and air. This reverence towards life is the basis of the Yogic teaching of *ahiṃsa*, or non-violence, non-injury, and non-harming.
>
> Today, the viability of earth's life systems is in danger. If humanity is to survive and thrive, we must learn to live in balance with nature. Now is the time to cleanse and heal the earth and to establish a sustainable relationship with the environment for generations to come.
>
> Therefore, as practitioners of Yoga we will:
>
> - Educate ourselves about the needs of the biosphere as a whole and our local ecosystems in particular.
> - Cultivate an appreciation for and conscious connection with the natural environments in which we live, including animals, plants, soil, water, and air.
> - Include care for the environment in our discussion of Yogic ethical practices.
> - Commit ourselves to policies, products, and actions that minimize environmental harm and maximize environmental benefit.
> - And if we are Yoga teachers or centers, we will incorporate these commitments into our work with students.

Many responders were enthused to connect around the convergence of Yoga and ecology. This gathering of enthusiasts and volunteers served as the founding nexus of the Green Yoga Association.

One initial project was raising awareness about the toxicity of the common Yoga mat, probably the most ubiquitous symbol of modern Yoga practice. The mat is seen frequently as a rolled up piece of plastic tucked under someone's arm as he or she walks to Yoga class. Plastics are classified according to their base ingredient, and unfortunately, the base ingredient for the vast majority of mats, polyvinyl chloride (also known as PVC or vinyl) is widely regarded as the most toxic.[2] Because of its high chlorine content, PVC creates dioxins and other toxins during both manufacture and destruction (by burning). Dioxin is a potent carcinogen that is accumulating widely across the planet today. It can be found as far away as the tissues of polar bears, and as close to home as in human breast milk. In addition, common softeners in PVC also cause health problems as they off-gas or rub off, being absorbed through human skin or breath.

The Green Yoga Association began to raise consciousness in the Yoga community on this issue in January of 2004, publicizing it at outreach booths at Yoga and environmental conferences, and later on our website and in our newsletter. Distributing one of the first non-toxic sticky mats in the U.S. complemented this work. Through our efforts, the U.S. mat market began to shift. Today, each major mat manufacturer in the U.S. offers at least one mat that is preferable to the highly toxic PVC. Awareness is growing, and many Yoga practitioners today know on at least a basic level that there is some problem with the common mat.

In working to green the Yoga studio, another familiar symbol of contemporary Yoga, the Green Yoga Association partnered with 70 such studios nationwide. These studios made many changes that might be found in other green business programs, such as minimizing paper, water, and energy use; recycling; encouraging students to bike or carpool; and reducing toxins through using non-toxic cleaning materials, bringing plants indoors, and choosing planet-healthy products for their boutiques.

In addition to the nuts and bolts of greening a business, the studios initiated special programs that expanded their role. For example, Yoga classes in Chicago and San Francisco were offered at botanical gardens, where students engaged in postures on the grass, learned about plant diversity, and helped with mulching and replanting. Others offered classes at the beach or partnered with environmental organizations. For example, World Yoga Center in Walnut Creek, California partnered with the non-profit organization Save Mount Diablo, raising money for this conservation organization, increasing public awareness of the mountain's endangered ecosystem, and teaching Yoga at Save Mount Diablo's events. Several Green Studios partnered with carbon offsets programs, supporting both local and global tree planting organizations. The studios communicated with each other via monthly phone meetings and through email.

In just three years from 2004 to 2007 the Green Yoga Association built two websites, produced two conferences, published four newsletters, distributed over 10,000 non-toxic Yoga mats, sponsored outreach booths at 15 Yoga and environmental conferences, and initiated two educational programs: the Green Studios Program and the Green Yoga Teacher Leadership Certificate Program. The Green Yoga Association continues to serve as a meeting place for Yoga teachers, studio owners, and practitioners with a strong concern for the health and well being of the environment. The Green Studios Handbook is online and various activities continue.[3]

While minimizing toxicity in mats and greening Yoga studios touch on important activist concerns, a perhaps even more profound shift comes from greening Yoga's actual practice. The collaborative group of Yoga teachers who experimented with strengthening the ecological aspects of Yoga practice, also considered the teachings of traditional Yoga. In observing over fifty experienced Yoga teachers bring their most creative and heart-felt efforts to greening their instruction, and in extended conversation with many of these people, I have seen certain patterns and themes emerge[4]. Experiments in practice tend to fall into common areas. Further, the practices demonstrate a desire to maintain the integrity of the Yoga tradition, while remaining open to the creative innovation needed to address our contemporary eco-crisis.

While it would be impossible to detail in this chapter all of the Green Yoga practices, highlights include: 1) offering devotional prayers, movements, chants, and blessings to all beings, including animals, plants, the five elements, and even the earth itself; 2) focusing attention on the subtle qualities of Yoga poses, including the ways they connect us to the elements of the planet and our body; to the gravitational field of the earth; to the life force (*praṇa*) that animates the universe; to the engaged experience of Yogic qualities of peace, courage, love, steadiness, non-violence, and strength; and to the soul qualities of the animals, plants, sun, moon, and other beings that many postures represent; 3) meditation on the five elements of earth, water, fire, air, and ether, either as an independent activity or as part of a Yoga class; 4) discussion and integration into one's life of the ecological implications of Yogic values, such as the *yamas* and *niyamas* of Patanjali's eight-limbed path; 5) engaging in eco-restoration or environmental protection as *Karma Yoga* (the Yoga of action); 6) meditation or *āsana* practice in an outdoor location, such as a forest grove, the beach, one's backyard, a city park or garden, or while on extended retreat in a wilderness area; 7) other outdoor practices such as walking meditation with relaxed sensory awareness[5] or sustained observation of a place or object in nature;[6] 8) telling the Story of the Universe through Yoga and dance movements, which reconnects Yoga to its embeddedness in universal processes and situates our current environmental choices as a new step in the cosmic-evolutionary path;[7] and 9) engaging in conscious community, such as the collaborative methodology of the dissertation research, the *Saṅgha* (community discussion) groups at the Green Yoga conferences, or the intentionally non-hierarchical structure of the Green Yoga Association council.

Through observing practices in action, I developed a model for eight paths of Green Yoga. The model provides a comprehensive framework for envisioning an ecologically-attuned Yoga and shows how Green Yoga grows out of the depth of Yoga's classical paths while also opening new avenues for spiritual practice aimed at healing our relationship with the world. Green Yoga as presented in this model is both old and new, traditional and creative.

In organizing Green Yoga practices, many fell naturally into traditional paths, such as *Jñāna*, *Bhakti*, *Haṭha*, *Rāja*, *Karma*, and *Tantra Yoga*s, while others did not. Thus, I created and named two new categories to fit these other areas: Āraṇyaka Yoga, or the Yoga of Going into the Forest, and Saṅgha Yoga, or the Yoga of Sacred Community. The two newly defined areas might have been squeezed into the other categories, but it seemed more consistent with the data to separate them. Cultivating conscious relationship with the forest and with the human community were arguably inherent in traditional paths; however, today we need to breathe new life into these practices by naming them explicitly. It is likely that these paths were not named earlier simply because they were assumed to be part of everyday living. Today, a conscious reengagement of these paths *as Yoga* is critical to our healing. The remainder of this chapter will describe each of the eight Green Yogas, and will conclude by offering a model for their interconnection.

Jñāna Yoga: Knowledge

Jñāna Yoga is the path of knowledge. In a classical context, it signifies developing the discriminative wisdom that enables one to distinguish between the ever-changing process of material nature, referred to as the field or object of awareness, and the constancy of the Self, or the witnessing consciousness. This discrimination brings the experience of steady peace, and a connection to the unchanging level of being that is our deeper nature.

Jñāna Yoga also encompasses the clear understanding that one underlying unity pervades, sustains, and makes possible all existence. In the tradition of the *Upaniṣad*s and the *Bhagavad Gītā*, the witnessing consciousness, or Self, was understood to be identical with the Self of all; the Self of the individual (*ātman)* was known to be the same as the Self of the universe (*brahman).* The oneness and pervasiveness of *brahman* is expressed clearly in this passage:

> With his hands and feet everywhere, with eyes, heads, and faces on all sides, with ears on all sides, He dwells in the world, enveloping all. . . . He is undivided (indivisible) and yet He seems to be divided among beings. He is to be known as supporting creatures, destroying them and creating them afresh. He is the Light of lights. . . seated in the hearts of all.
> *Bhagavad Gītā* XIII: 13–17

Contemporary Green Yogis resonate strongly with the teaching that divinity is present in all of material nature, from the birds and bees to the oceans and minerals, and even including the wider cosmos. Further, the interconnectedness this idea expresses is strongly akin to contemporary ecological wisdom.

The Yoga tradition emphasizes interconnection not only on a spiritual but also a physical level, predating the concept of the web of life. Humans are understood to be linked with the cosmos in a sacrificial relationship of give and take, expressed most directly through the processes of eating and breathing. The *Taittiriya Upaniṣad* states that everything is to be looked upon as food, or the substance that ultimately gives life to another; this *Upaniṣad* further details that one who achieves such right vision lives in ecstasy and bliss (*TU* II:21; III:10.6–7).

Green Yoga practices enable one to experience the interconnection of being while also strengthening the power of discriminative wisdom. Meditation on the elements and the various forms they take aides the practitioner in perceiving the most basic of interconnections. For example, the physical body is part of the earth's food and soil cycle, which in turn receives its energy and impetus from the sun. The water in our bodies is part of the larger water cycle of the planet. The air we inhale and exhale is only a small part of larger atmospheric processes. Through *āsana* practice and guided meditation, breath and body may be sensed literally as a subset of earth.

Green Yoga highlights the life systems of the body as an integral part of the life systems of the earth. Human health is inextricably linked with the health of the earth The right understanding of life is supported and guided by a sense of interconnected and underlying unity, necessitating a close consideration of human conduct and its impacts on other beings. As one develops discriminative wisdom, one knows that this interconnection is the unchanging reality, the deepest aspect of our being and that of the planet. This awareness provides a foundational wisdom that supports the other Yogas.

Bhakti Yoga: Reverence

Bhakti Yoga is the conscious practice of reverence or devotion. In Green Yoga, the reverence for and intimacy with nature that was always part of the Yoga tradition becomes conscious and explicit and is translated into contemporary terms. Elements from nature such as mountains, rivers, trees, lions, snakes, monkeys, birds, bees, flowers, the sun and the moon have all been worshiped in Yoga, and are celebrated in art, myth, drama, and in Yoga postures and salutations. The universe itself, and all within it, is viewed as the body of God. Green Yoga recovers this historic and profound reverence for nature, integrating it into each Yoga experience.

Haṭha Yoga classes begin very close to home by practicing self-love and by sending blessings for bodily healing. The actual practice of *Haṭha Yoga* fosters reverence for breath and for the awakening, embodied life force within. As one becomes aware of the gifts of the body, one experiences awe at the mysteries of all life. Then, because reverence is a mind-set that colors all perceptions, feelings of compassion toward the body gradually expand to include other beings. Through learning to treat the body tenderly, one grows to experience tenderness also towards animals, plants, soil, water, and air.

Many Yoga practitioners today are familiar with the Sun Salutation, and some also with the Moon Salutation. A Green Yogi is likely to give gratitude to the sun and moon while practicing their respective salutations. Again, developing this consciousness of gratitude and awe prepares one to view the actual sun and moon with awe and reverence. Devotional poems to the earth, centering meditations focused on reverence, or a *bhakti*-oriented story told during class also foster a reverential mind-set. Gradually, Yoga becomes a means of experiencing the miraculousness of life, rather than simply a way to generate fitness.

As the consciousness of reverence expands, the practitioner ultimately perceives the earth and cosmos as sacred. Since the time of the Vedas, Yoga has conceived of the cosmos as the body of God. The earth has been conceived of as goddess, who fed and nurtured all creatures, and upon whom devotees could call for understanding and support. Contemporary Yoga in the United States has embraced the idea of the body as a temple, but has not clearly spoken about the earth body as sacred. Human consciousness has now expanded to be able to conceive of the planet in its entirety; Green Yoga cultivates reverence and devotion to this sacred whole.

Recovering a reverential attitude is essential if we are to regain our sanity in relation to the earth. Reverence for nature is an essential ingredient of earth care and protection. Green Yoga fosters awe and wonder at the miracle of life on this planet, and at the majesty of the sun, moon, and stars. Ultimately, the love that is engendered through this devotional practice purifies and heals both the practitioner and the earth itself.

Āraṇyaka Yoga: Going into the Forest

Āraṇyaka Yoga is the path of connecting with the soul of the earth and cosmos through spending time in nature. So much of contemporary life separates human beings from nature, and Green Yoga seeks to heal this fundamental wounding, building in its place a conscious and living relationship with the world. I derived the name of this Yoga, or what I call *The Yoga of Going into the Forest*, from the Sanskrit word *āraṇyakam*, meaning forest. It invokes the ancient sense of the sanctity of wild places, far from the domesticating influence of cities. The presence of nature is always accessible, as even the most man-made materials have their origins deep in the life of the earth and sun. In most cases, however, the term Āraṇyaka Yoga refers to the act of going outside, actually getting *out* into nature.

A basic practice of this Yoga is spending time in one particular place in nature, and getting to know it intimately. Some Green Yogis find a place near their home that has access to water such as a river, stream, bay or ocean. Some find a special connection with a nearby tree or park, or a mountain or hill they visit regularly. Regular meditation on the actual rising of the sun is also a practice of Āraṇyaka Yoga. Sunrise meditation connects one with the turning of the planet, with the warm generosity of the sun, and with the peacefulness of the early morning. The *Gāyatrī Mantra*, an ancient Vedic prayer dedicated to the light of the cosmos, or some other mantra might be invoked. Some Green Yogis celebrate solstices, equinoxes, and full moon events with extensive outdoor rituals. The ancient practice of spiritual pilgrimage, in which one walks for an extended period of time in a natural area, is also a Green Yoga practice.

Green Yogis form daily habits that connect them with nature close to home; at the same time more extended visits on a seasonal or annual basis make it possible for those who live in cities to experience the wildness of undomesticated nature. Yoga retreats in forests, near oceans, or other natural places provide deep soul rest. The healing of our soul-estrangement from nature cannot take place on a purely metaphoric level, but requires actual contact with trees, moon, sun, stars, ponds, forests, and oceans. Such re-connection is actually a way back to the inner self, and back to the deepest guidance found in the heart and the world.

Haṭha Yoga: Sacred Embodiment

Haṭha Yoga is the path of fully embracing spirituality through the vehicle of the body. As most commonly practiced in the United States today, *Haṭha Yoga* consists primarily of *āsana*s (postures) and *prāṇāyāma* (breathing exercises). Through connecting with the body, the practice of *Haṭha Yoga* provides an immediate, sensate link to nature, to the self as a living, breathing, fleshy being. While in daily life some might forget the physical body and unwittingly experience themselves solely as thinking heads, in *Haṭha Yoga* this becomes impossible. In this habitation of the body is the connection to other animals, plants, and the minerals of the earth, water, and air.

For example, in the embodied experience it becomes possible to actually sense the elements of the earth running through the body as described in earlier sections, and an ecologically-attuned teacher is likely to highlight these connections. Through the various movements and heightened awareness that arises during Yoga practice, one can feel the water pooling in the mouth and the blood running through the veins. This can guide awareness to the recognition of the waters of the earth that run through the body. *Prāṇāyāma* becomes an opportunity to recognize inter-being within the atmosphere of the planet, and with all other living beings, both plants and animals, that also partake of the same atmosphere. A sensitive teacher can guide a student's understanding here, helping one to experience on a sensory level how the systems of the body are but a small subset of the systems of the earth.

Also, consciousness in general becomes more pliable during the practice of *Haṭha Yoga.* Relaxing, breathing, and focusing on sensations leads to mental as well as physical flexibility, creating an opening beyond the rigidity of the ordinary, thinking mind. In this liminal space, new soul perceptions are accessible. In Green Yoga, one may experience a soul resonance with the animals, plants, and other beings whose forms one imitates. While on a literal level, the shape of the pose may not necessarily connect with its namesake, when the pose is experienced playfully, poetically, or on the level of qualities, the connection becomes clear. For example, Sun Salutation expresses the brilliant radiance of the sun and gratitude for its gifts. Cobra Pose embodies the earthiness, beauty, and mystery of the cobra. During the course of a *Haṭha Yoga* practice session one might feel a soul resonance with the sun, moon, fish, cobra, trees, lotus flower, bees, dolphin, and many other beings. The soul is stretched and expanded as to embody relatedness with other beings through the poses.

Attending to inter-being with other life forms and the body of the planet as an integrated part of *Haṭha Yoga* practice encourages vast openings towards oneness with the Earth. One can perceive sacred connections between body and cosmos, and experience the body itself as a sacred ecological mode. One comes to know the self not as separate from but part of. In all of these ways *Haṭha Yoga* helps to release self-perception as other than, superior to, or separate from nature. Yoga gradually releases the anthropocentric mode and radically contextualizes humanity within the cosmic and biotic spheres.

***Rāja Yoga:* Conscious Evolution**

Rāja Yoga is the path of comprehensive self-transformation. Literally *the royal path*, *Rāja Yoga* was best described by Patañjali as the eight-limbed path of Yogic evolution. *Rāja Yoga* encompasses a broad range of practices from ethical commitments to breath enhancement to full meditative absorption; all are intended to purify and stabilize the mind. Individual self-transformation is a critical step in earth healing. In Green Yoga, working to purify the body-mind continuum is an act of compassion undertaken for the whole world. The effort at self-transformation to live in harmony with the earth contributes to the positive evolution of the human species, so needed at this time in history.

A fundamental aspect of working with the mind is commitment to the *yamas* and *niyamas*, the ethical foundation of *Rāja Yoga*. For example, in applying *ahiṃsā*, nonviolence, one looks clearly at the effects of actions, and begin to make choices that cause less harm. One might choose to reduce dramatically plastics consumption in order to minimize harm to the oceans. Through awareness of *satya*, or truth telling, one recognizes that denial of the damage being done to the planet takes one away from reality. One begins to break through denial to see the facts as they are, a root necessity for any seeker of truth. Through contemplation of *asteya*, the principle of non-stealing, one recognizes that consuming more than is needed is a form of stealing from the earth. To avoid stealing trees from the forests, a person may choose to use less paper; to avoid stealing habitat and life-giving liquid from the rivers, one chooses to use less water.

Through application of *brahmacharya*, or conscious energy management, one becomes aware that seeking pleasures through excessive travel and consumerism depletes the earth and human alike. A person may then seek to find pleasures that are closer to home and less taxing on the earth, and to transmute pleasure-seeking itself into an overflowing love of the world. Through awareness of *aparigraha*, or non-grasping, one recognizes that the fear-based behavior of hoarding denies the giving-ness of the cosmos. Instead the yogi practices generosity, sharing and cooperating with one's neighbors and the earth for global security. Similarly, the *niyamas* are conscious practices for cultivating positive qualities that support earth care.

Stabilization of the body through conscious sitting and other postures (*āsana*) helps to still the mind, strengthening and purifying it. Expansion and conscious drinking in of the breath (*prāṇāyāma*) saturates every cell with life force energy. Green Yoga works with the senses (*pratyāhāra*) by experiencing the world with relaxed sensory awareness and by non-attachment to things. The practitioner releases the actions of the senses from the distortions of the grasping mind. One takes time to sit quietly and feel divinity within, practicing total withdrawal of the senses. All of these stages profoundly provide support on the path of conscious transformation.

Extended concentration (*dhāraṇā*) on a chosen object enables one to penetrate into deeper levels of being. In Green Yoga one might concentrate on an element of nature, on the rising sun, or perhaps on the beating of one's heart. *Dhāraṇā* is used as a means of connecting on deep levels with the uniting presence that animates all aspects of the earth and cosmos. It is in meditation (*dhyāna*) and ultimate absorption (*samādhi*) that the individual rests in imperturbable peace of mind. The practitioner has moved beyond the dualities of pleasure or pain, good or bad, and instead views all with equanimity.

While traditional interpretations of Patañjalian Yoga may indicate that in the final stages of absorption one is completely separate from the world, in this ecologically-attuned Yoga one remains fully engaged in it. One experiences everything with ecstasy and is able to act on behalf of all beings. Green Yoga understands meditation and absorption as not requiring or leading to diminished participation in life, a tenet in line with some scholars' interpretation of Patañjali's meaning, such as Ian Whicher expresses in chapter three of this volume.

The evolution of individuals practicing Yoga supports the evolution of the human species. Green Yoga expands one's perspective from transformation of the individual into transformation of the species. The spiritual evolution of each one contributes to the spiritual evolution of the whole. The work towards becoming more conscious and loving beings is a gift of consciousness to the whole planet.

Karma Yoga: Action

Karma Yoga is the path of selfless service and a primary Yogic path of spiritual development. The process of reaching out to the Yoga community with the Green Yoga Values Statement, educating the public about Yoga mats and distributing non-toxic mats, producing conferences, and all other actions undertaken by the Green Yoga Association may be considered as falling under the rubric of *Karma Yoga*. Given the extent of the environmental crisis, being clear on how to act skillfully in raising consciousness and caring for the planet is very, very important.

The *Bhagavad Gītā* teaches that one should not undertake action with the intention of gaining a certain result, but rather with the intention of service. Krishna says "Do the work that comes to you—but don't look for the results. Don't be motivated by the fruits of your actions" (*BG* II:47), and elsewhere "Strive constantly to serve the welfare of the world; by devotion to selfless work a man attains the supreme goal of life" (*BG* III:19). For many of us, this principle is counterintuitive, as people tend to believe that the only reason to undertake an action is to achieve a given result. However, attachment to results binds an individual, creating karmic links that actually prevent an openhearted and completely free engagement in the action.

When discussing actions intended to heal the planet, one of the first questions that arises is whether or not it is too late. "Can the planet be saved?" Green Yoga answers this question by saying that ultimately the outcome of our actions doesn't matter. As a duty and privilege compassionate human beings cherish the earth. One acts not out of fear, anger, or guilt, but out of love. Dharma heals and cares for the earth; engaging it brings peace and joy. In Green Yoga, action in service of the earth becomes a form of self-healing; through caring for the earth, the soul connection with the earth is healed, and in this way, the person is transformed. Yoga as service in this way becomes a path of transformation of self; both self and world are healed through such action.

Another key principle is that action is about working on oneself first. This means that it is *not* about trying to get others to change. Mahatma Gandhi, one of the great karma yogis of all time, did not focus first on getting others to change. Rather, his strongest attention was always on self-purification, on self-change. In his own ashram, Gandhi could require high standards of ethical action, because he himself upheld them. And from his example, hundreds of thousands of people worldwide have been inspired to become like him.

Thus, action intended to heal the world begins with transforming one's own lives and habits. The Green Yogi commits to causing the earth less harm and to seeking ways to give back some of the blessings we have received from it. As described in the section on *Rāja Yoga*, one might make a commitment to minimizing harm to forests, and thus reduce the use of virgin paper. One might commit to minimizing harm to oceans, and thus significantly reduce the use of plastics. This is the fundamental practice of *ahiṃsā*, or nonviolence, that was discussed in the previous section. Through dedicating oneself to becoming more compassionate and peaceful, one becomes strengthened and transformed, primed for wider action in the world.

In contemplating reaching beyond a mundane life to raise consciousness with others, Green Yoga helps one to be peaceful with activist energy. *Haṭha Yoga* and meditation are calming and grounding. Each Yoga class becomes a mini-retreat where one practices surrendering results. As one lets go of the outcome or form of each pose, one prepares to similarly relax while acting in the world. In the warrior and other standing poses, one embodies the courage, discipline, and engagement of the archetypal warrior; Green Yoga adds to this repertoire the peacemaker's powerful love and truth-telling.

While many may consider reaching out to be the first step in action, instead, action happens naturally when one feels called, and occurs in balance with an extensive process of self-healing, self-transformation, and reflective retreat. Withdrawal for retreat helps focus energy and gives needed insight. Reflection is both a necessary preliminary and a necessary counterbalance to action. Without ongoing reflection, action becomes empty and forced. Journal writing, walks in nature, meeting in supportive groups, or inner dialogue with the Divine presence can provide nourishing connection with intuition, soul, and sacred guidance.

Action on behalf of the earth is the right path, the dharma, for compassionate human beings. One acts out of a desire to serve the sacred through loving and caring for this beautiful planet. And the practices of Green Yoga bring strength to act with non-attachment to outcome. Walking the path of action means not seeking to change others, and yet working to increase awareness of how people could all love the planet better. It means acting as skillfully as possible to create positive results, and yet releasing attachment to those results.

Saṅgha Yoga: Community

Saṅgha Yoga is the path of evolving spiritually through the vehicle of sacred community. Saṅgha Yoga forms bonds of connection and love that nourish and strengthen and that serve as sacred teachers. Green Yogis recognize that isolated practice is not sufficient to heal today's world and are challenged to recognize unity with others despite differences.

The current era is coming to recognize the path of relationship and conscious communication as a sacred Yoga. Rather than the archetype of withdrawal as an ascetical hermit, today's Yogi or Yogini is an engaged being who includes caring connection with others and the earth as part of the path of internal alchemy. For many, intimate partnership is a path to expanded spirituality. While those engaged in Green Yoga may enjoy solitary immersion in nature as soul-restoring and deeply enriching (as in Āraṇyaka Yoga), they are also aware that reaching out to other humans is an essential practice for healing today's world, and work to balance the poles of retreat and engagement.

Saṅgha Yoga is essential to an ecological Yoga because the group helps to awaken one from the anti-ecological stupor of mainstream culture. Contemporary culture is antithetical to environmental appreciation and respect for a balanced, ecological lifestyle. The changes needed to balance human lives with the needs of the earth are not changes can be made alone. The support and cooperation of a group of like-minded individuals is essential in transforming cultural patterns.

In individualism many attributes that are detrimental to the health of the environment, such as greed and fear, find fertile breeding ground; community is the antidote to individualism. In the difficult but rewarding work of Saṅgha one may find affirmation, support, encouragement, insight, challenge, and respectful difference. Conflicts inevitably arise, and learning to work with and through them provides needed impetus for our growth. While solitary practice requires nonjudgmental self-awareness, group practice and interaction demand it perhaps to an even greater degree.

Like the other Yogas, Saṅgha involves methods of awareness and requires disciplined attention. Receptive listening; inner honesty; respectful sharing; and compassionate observation of self, others and the group are all essential. Saṅgha Yoga practices can include engaging in Yoga practice together, establishing and maintaining clear ground rules and commitments, listening and sharing, communicating difficult feelings and emotions, sharing leadership and decision making, coming to agreement on significant actions, and participating in joint projects.

In community one experiences the variety and beauty that are present in nature. The individuality of each group member mirrors the individuality of each tree or blade of grass. The healthy diversity found in community challenges and enriches experience, reflecting the diversity of a healthy ecosystem. Ultimately, through the practice of Saṅgha, one becomes aware of the oneness of Being; while people are different and distinct from every other individual, an undeniable unity and presence becomes known. While at times it might be more comfortable to engage in spiritual practice by oneself,

Green Yoga recognizes that this is not sufficient today. The age of individual realization has passed. To heal the world requires the fire of living community.

Tantra Yoga: Embracing the Unity of Opposites

Tantrism is arguably the most ecological of Hindu philosophies, and is the root from which *Haṭha Yoga* developed. It was Tantrism that most clearly pronounced the divinity pulsing in every aspect of nature. Further, Tantrism affirmed the possibility of achieving spiritual liberation through the vehicle of the world, rather than needing to pull away from or reject it. Literally, *tantra* means weaving together, and Tantrism has always been associated with embracing and unifying the opposites, including male and female, spirit and nature, light and dark, grief and joy.

Of key importance for an ecological Yoga, Tantrism valorizes the phenomenal world, rather than lifting up the spiritual realm alone. While other philosophies of Hinduism pointed to the divine presence in all of nature, they also demonstrated an ambivalence about materiality that diminished its significance and tended towards other-worldliness. Nature was seen as feminine, and both were devalued in relation to transcendent consciousness, seen as masculine. Tantrism subverts these categories by revalorizing both material nature and the feminine, honoring the immanent *and* the transcendent. This has major resonance for today's Green Yogis, who strongly desire spiritual growth and transformation, but are also concerned with embracing physicality and caring for the phenomenal world.

In opening to the opposites, today's Green Yogis seek to be receptive to both light and dark in relation to ecology. While one may experience joy and peace during time spent in nature, one is also aware of deep grief and concern for ecological crisis. Embracing both poles is an aspect of Green Yoga. The light that is present in the world guides human actions toward healing and upliftment for the planet. The wisdom of this light may be invoked through a variety of means, such as quiet meditation; basic awareness of cellular vitality; specific prayers, such as *Gāyatrī Mantra*; or visual concentration on fire such as a campfire, a candle, or the sunrise. Similarly, opening to the dark brings clarity and compassion. Ordinary lives do not give space for the deep grief many people today experience at the ecological devastation our time is witnessing. Green Yogis seek to consciously name and grieve specific losses, either individually or collectively, and to acknowledge the universality of suffering. As people become aware of suffering, compassion can arise, along with the desire to minimize suffering. A Green Yogi may send blessings to the dark places of her own suffering and the dark places in the suffering of others. Such a yogi acknowledges sorrow at the ways in which we are not living in balance with the needs of the planet.

Spiritual ecology is a contemporary pathway into Tantrism. Green Yogis strongly valorize materiality while also seeking transcendent consciousness. The body of this earth is sacred, and every aspect of human life serves as food for the journey. A Green Yogi embraces the highs and the lows, the grief and the joy. Light and dark can exist simultaneously.

An Integrated Model for Eight Paths of Green Yoga

This chapter has shown how each of the traditional Yogas may be understood in the context of Green Yoga. In addition, this model goes beyond traditional Yoga in naming two new paths as medicine for our age: Āraṇyaka and Saṅgha Yogas. These correct significant cultural imbalances that prevent ecological living: isolation from the natural world and isolation from each other.

The Yogas may also be understood as fitting into one cohesive framework for Green Yoga (see Figure 1). Knowledge of interconnection (*Jñāna Yoga*) and a reverential mindset (*Bhakti Yoga*) form the foundational entry point from which to move into the other Yogas. As the process of self-transformation becomes well established through going into the forest (*Āraṇyaka Yoga*), sacred embodiment (*Haṭha Yoga*), and transformation of mind-states (*Rāja Yoga*), action to heal the world (*Karma Yoga*) becomes a natural and joyful out-flowing. This action and the resultant learning in turn deepen awareness of interconnection and further strengthen our gratitude and awe. In this way, the cycle continues in an ongoing flow of deepening awareness, reverence, self-transformation, and action.

In this model the pole of retreat (in which one develops discriminative wisdom and fosters a reverential mindset) flows into the pole of engagement (in the form of healing the world) and vice versa. Self-transformation mediates and is facilitated by the learning of each pole. Self-transformation is also the outcome of the interplay between retreat and engagement. Right knowledge, reverence, self-transformation, and action are all held within the container of the sacred community (*Saṅgha Yoga*), which supports the growth and learning of the other Yogas. Finally, realization of unity consciousness (*Tantra Yoga*) grows out of the sum of all the other practices.

What is needed today is a fresh perspective. We need to find our way out of the cultural patterns that are hurting this planet. The powerful practice of Yoga can help us to remake ourselves on all levels of our being, and to consciously shift our cultural patterns. Through living our Yoga in this way, we become vehicles for transformation in the world, and vehicles for the transformation of Yoga practice and teaching itself.

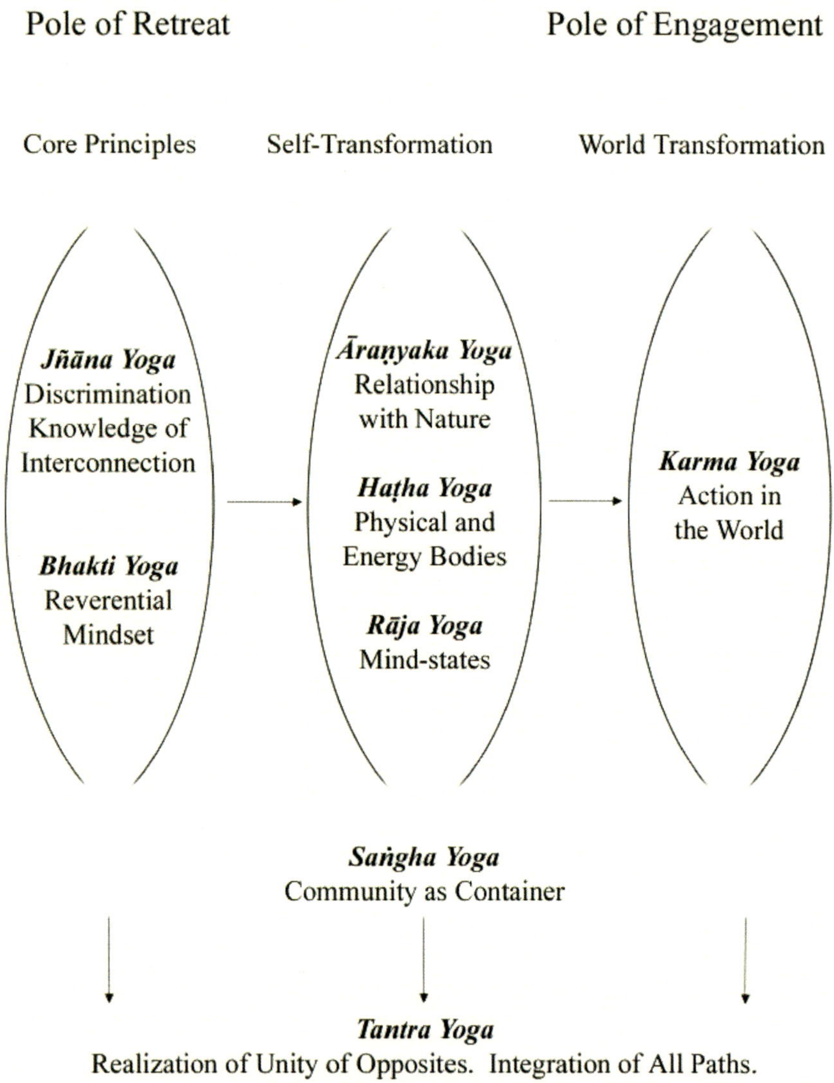

Figure 1. Green Yoga Model.

Endnotes

[1] Kripalu Yoga is a style of Modern Postural Yoga that includes a strong emphasis on devotion (*Bhakti*), attention to the sensation and flow of *praṇa* (life force) while holding the poses, and a spontaneous posture flow, also known as meditation-in-motion.

[2] For a more complete analysis of polyvinyl chloride and the problems with the common Yoga mat, see "Rethinking Yoga Mats: The Search for a Green Solution." *Green Yoga Times* (2004/2005A, Fall/Winter), I(1), 8, 9 .

[3] For more information, visit www.greenyoga.org.

[4] Those who have been most influential in my observations and learning regarding Green Yoga include the members of my dissertation collaborative: Bob Bruce, Tanuja Pat Daniel, Ben Lord, Agi Hasita Nadai, and Leanne Ovalles; the members of the Green Yoga Teacher Leadership Instructional Team: Christopher Key Chapple, Gillian Kapteyn Comstock, Russell Comstock, and Agi Hasita Nadai; and the members of the extended study course of the Teacher Leadership Program: Andy Amend, Nicole Cammaert, Jasmine Lieb, Joyce Eden, Guy Gabriel, David Lurey, Masaaki Nakajima.

[5] Related to the Yogic practice of pilgrimage and the Native American tradition of fox walking.

[6] Related to the Yogic practice of nature contemplation and the Native American tradition of Good Medicine Place.

[7] This practice was developed primarily by Agi Hasita Nadai, who refers to it as YogaGaia.

BIBLIOGRAPHY

Abram, David. *The Spell of the Sensuous: Perception and Language in a More-Than-Human World* (1st Ed.). New York: Pantheon Books, 1988.

Adwayananda, Sri (K. Padmanabha Menon). *Atmaswarupam: One's Own Real Nature.* Austin, Texas: Advaita Publishers, 1988.

Agarwal, Anil. "Human-Nature Interactions in a Third World Country." In George James, Ed., *Ethical Perspectives on Environmental Issues in India.* New Delhi: A.P.H. Publishing Corporation, 1999, 31–72.

Akula, Vikram K. "Grassroots Environmental Resistance in India." In *Ecological Resistance Movements: The Global Emergence of Radical and Popular Environmentalism,* Ed. Bron Raymond Talyor, 127–45. Albany: State University of New York Press, 1995.

Alley, Kelly D. "Ganga and Gandagi: Interpretations of Pollution and Waste in Banaras." *Ethology* 33 (Spring 1994): 127–45.

Alter, Joseph S. *Yoga in Modern India.* New Jersey: Princeton University Press, 2004.

Anantananda, Swami (Guest Ed.) (March 1990). *Nature: The Face of God. Darshan Magazine.* South Fallsburg, NY: Siddha Yoga Dham Association, 1990.

Apfell-Marglin, Frederique. "Gender and the Unitary Self: Looking for the Subaltern in Coastal Orissa." *South Asian Research,* 1995.

Āraṇya, Hariharānanda. *Yoga Philosophy of Patañjali*, Trans P. N. Mukerji Albany: State University of New York Press, 1983.

Āraṇya, Hariharānanda. *The Sāṃkhya-sūtras of Pañcaśikha and The Sāṃhyatattvāloka*. Delhi: Motilal Banarsidass, 1977.

Arya, U. *Yoga-Sūtras of Patañjali with the Exposition of Vyāsa: A Translation and Commentary – Vol. 1: Samādhi-Pāda*. Honesdale, PA: Himalayan International Institute, 1986.

Aurobindo, Sri. *The Secret of the Veda*. Pondicherry: Sri Aurobindo Ashram, 2004.

Baxter, Devas Ken. KYTA Conference: "The Quest for An Ecological Yoga," in *Kripalu Yoga Teachers Association Yoga Bulletin*, 5(2), 1, 6.

Banwari. *Pancavati: Indian Approach to Environment*. Trans. Asha Vohra. Delhi: Shri Vinayaka Publications, 1992.

Bhabha, Homi K. *Anish Kapoor*, Hayward Gallery and University of California Press, 1998.

Bose, Abinash Chandra, Trans. and Ed. *Hymns from the Vedas*. Bombay: Asia Publishing House, 1966.

Brooks, Charles R. *The Hare Krishnas in India*. Princeton: Princeton University Press, 1989.

Callahan, Daren. *Yoga: An Annotated Bibliography of Works in English, 1981–2005*. North Carolina: McFarland & Company, Inc., 2007.

Callicott, J. Baird, and Roger T. Ames, Eds. *Nature in Asian Traditions of Thought: Essays in Environmental Philosophy.* Albany: State University of New York Press, 1989.

Callicott, J. Baird. *Earth's Insights.* Berkeley and Los Angeles: University of California Press, 1994.

Carpenter, David. *Yoga: The Indian Tradition,* edited by Ian Whicher and David Carpenter. New York: RoutledgeCurzon, 2003.

Chand, Devi, Ed. *The Atharvaveda*. Delhi: Munshiram Manoharlal, 2002.

Chapple, Christopher K., and Evelyn Tucker, Eds. *Hinduism and Ecology: The Intersection of Earth, Sky and Water.* Cambridge: Harvard University Press, 2000.

Chapple, Christopher Key, Ed. *Ecological Prospects: Scientific, Religious, and Aesthetic Perspectives.* Albany: State University of New York Press, 1994.

———. *Jainism and Ecology*: Nonviolence In the Web of Life. Cambridge, Mass.: Harvard University Press, 2002.

Chapple, Christopher Key. "Hinduism and Ecology." In Tucker and Grim, Eds., *Worldviews and Ecology: Religion, Philosophy, and the Environment.* Maryknoll, N.Y.: Orbis Books, 1994.

———. "India's Earth Consciousness," In Michael Tobias & Georgianne Cowan, (Eds.), *The Soul of Nature* (pp. 145–151). New York: Continuum, 1994.

———. "Hinduism and Deep Ecology," In David Landis Barnhill & Roger S. Gottlieb, Eds., *Deep Ecology and World Religions*, pp. 59–76. Albany: State University of New York Press, 2000.

———. "Yoga," in *The Encyclopedia of Religion, Culture and Nature*. New York: Thames Continuum, pp. 1782–1785, 2005.

———. *Nonviolence to Animals, Earth, and Self in Asian Traditions*. New York: State University of New York Press, 1993.

———. *Yoga and the Luminous*. Albany: State University of New York Press, 2008.

Clarke, J. J. *Oriental Enlightenment: The Encounter between Asian and Western Thought*. London and New York: Routledge, 1997.

Cornell, Laura. "Green Yoga: A Collaborative Inquiry Among a Group of Yoga Teachers." Doctoral Dissertation, California Institute of Integral Studies, 2006.

———. "What Can Yoga Contribute to the Environmental Movement?" in *Green Yoga: The Newsletter of the Green Yoga Association*. Volume I, Issue 2, 2006, p. 3.

Coward, Harold. "The Ecological Implications of Karma Theory." In Nelson, Ed., *Purifying the Earthly Body of God*. Albany: State Univerity of New York Press, 1998.

———, Ed. *Population, Consumption, and the Enviornment: Religious and Secular Responces*. Albany: State University of New York Press, 1995.

———. *Visions of a New Earth: Religious Perspectives on Population, Consumption and Ecology.* Albany: State University of New York Press, 2000.

Crawford, S. Cromwell. *The Evolution of Hindu Ethical Ideals.* Delhi: Arnold-Heinemann, 1984.

Cremo, Michael A., and Mukunda Goswami. *Divine Nature: A Spiritual Perspective on the Environmental Crisis.* Los Angeles: Bhaktivedanta Institute, 1995.

Dasgupta, Surendranath. *Yoga As Philosophy and Religion.* Delhi: Motilal Banarsidaas, 1978.

Desai, Yogi Amrit. *Kripalu Yoga: Meditation in Motion: Book 1* (Rev. Ed.). Lennox, MA: Kripalu Yoga Fellowship.

Deutsch, Eliot. "A Metaphysical Grounding for Natural Reverence: East-West." In Callicott and Ames, Eds., *Nature in Asian Traditions of Thought.*

———. "Vedanta and Ecology." *Indian Philosophical Annual* (Madras) 7 (1970): 79–88.

Diez, Jordi and O. P. Dwivedi, Eds. *Global Enviornmental Challenges.* Toronto and Ontario: Broadview Press, 2008.

Dviveda, Vrajavallabha. "Having Become a God, He Should Sacrifice to the Gods," in *Ritual and Speculation in Early Tantrism: Studies in Honor of André Padoux*, Teun Goudriaan, Ed. Albany: SUNY Press, 1992.

Dwivedi, O. P. "Dharmic Ecology," In Christopher Key Chapple & Mary Evelyn Tucker, Eds., *Hinduism and Ecology: The Intersection of Earth, Sky, and Water,* pp. 3–22). Cambridge: Harvard University Press, 2000.

———. *Darshan: Nature and the Face of God.* Vol. 36 New York: S.Y.D.A. Foundation, 1990.

———. "Environmental Protection in the Hindu Tradition." In James, Ed., *Ethical Perspectives.* New Delhi: A.P.H Pub. Corp., 1999, 161–88.

———. "Satyagraha for Conservation: Awakening the Spirit of Hinduism." In Gottlieb, Roger S., Ed., *This Sacred Earth: Religion, Nature, Environment.* New York: Routledge, 1996.

Dwivedi, O. P., and B. N. Tiwari. *Environmental Crisis and Hindu Religion.* New Delhi: Gitanjali Publishing House, 1987.

Eliade, Mircea. *Shamanism: Archaic Techniques of Ecstasy.* Translated by Willard R. Trask. Princeton: Princeton University Press, 1963.

———. *Yoga: Immortality and Freedom.* Trans. by Willard R. Trask. 2nd Edition, Bollingen Series No. 56. Princeton: Princeton University Press, 1969

——— *Patañjali and Yoga.* Trans. by Charles Lam Markmann. New York: Schocken Books, 1975.

Ferrer, Jorge. *Revisioning Transpersonal Theory: A Participatory Vision of Human Spirituality.* Albany: State University of New York Press, 2002.

Feuerstein, Georg. *The Yoga-Sūtra of Patañjali: A New Translation and Commentary.* Folkstone, England: Wm. Dawson and Sons, Ltd., 1979.

———. *The Yoga Tradition: Its History, Literature, Philosophy, and Practice.* Prescott, Arizona: Hohm Press, 1998.

———. *Green Yoga.* Eastend, SK: Traditional Yoga Studies, 2007.

———. *Yoga Morality: Ancient Teachings at a Time of Global Crisis.* Hohm Press, 2007.

Flood, Gavin. *Body and Cosmology in Kashmir Śavism.* Lewiston: The Edwin Mellen Press, 1993.

Frawley, David. *Yoga and the Sacred Fire: Self-Realization and Planetary Transformation.* India: Motilal Banarsidaas Publishers, 2006

Gheranda. *The Gheranda Samhita.* Tranlsated by Rai Bahadur Srisa Chandra Vasu. Delhi: Sri Satguru Publications. 1979. First published, 1914–15.

Gordon White, David. "Yoga in Early Hindu Tantra." In *Yoga: The Indian Tradition,* Edited by Ian Whicher and David Carpenter, 143–161. New York: RoutledgeCurzon, 2003.

Gosling, David L. *Religion and Ecology In India and Southeast Asia.* London and New York: Routledge, 2001.

Gottlieb, Roger S., Ed. *This Sacred Earth: Religion, Nature, and Environment.* New York: Routledge, 1996.

Griffith, Ralph T. H., Trans. *The Hymns of the Rgveda.* Delhi: Motilal Banarsidass, 1973.

Gupta, Sanjukta. "The Maṇḍala as an Image of Man," in Richard Gombrich, Ed., *Indian Ritual and its Exegesis*. Delhi: Oxford University Press, 1988, p. 32–41.

Haberman, David L. *River of Love in an Age of Pollution: The Yamuna River of Northern India*. Berkeley: University of California Press, 2006.

Halbfass, Wilhelm. *Tradition and Reflection: Explorations in Indian Thought*. Albany: State University of New York Press, 1991.

—— *On Being and What There Is: Classical Vaiśeṣika and the History of Indian Ontology*. Albany: State University of New York Press, 1992.

Halifax, Joan. *Shaman*. London: Thames & Hudson, 1982.

Jacobsen, Knut A. 'What Similes in Sāṃkhya Do: A Comparison of the Similes in the Sāṃkhya texts of the Mahābhārata, the Sāṃkhyakārikā and the Sāṃkhyasūtra.' *Journal of Indian Philosophy* 34 (2006): 587–605.

———. 'BhagavadGītā, Ecosophy T and Deep Ecology.' In *Beneath the Surface: Critical Essays in the Philosophy of Deep Ecology*, Ed. Eric Katz, Andrew Light, and David Rothenberg, 231–52. Cambridge: The MIT Press, 2000.

———. *Prakṛti in Sāṃkhya-Yoga: Material Principle, Religious Experience, Ethical Implications*. New York: Peter Lang, 1999.

———. "Bhagavad Gītā, Ecosophy T and Deep Ecology." In *Inquiry: An Interdisciplinary Journal of Philsophy and the Social Sciences* Vol. 39 (June 1996): pp. 233–234.

Jain, Dr. Kamala. "Consumerism, Environment and Non-Possessiveness," in *Jain Study Circular,* 11–14, 2000.

James, George A. *Ethical Perspectives On Environmental Issues In India.* New Delhi: A.P.H. Publishing Corporation, 1999.

Janakiraman, Yogacharya, and Carolina Roso Cicogna. *Solar Yoga.* New Delhi: Allied Publishers Limited, 1989.

Jayakar, Pupul. *The Earthen Drum.* New Delhi: National Museum, 1980.

——. *The Earth Mother: Legends, Ritual Arts, and Goddesses of India.* New York: Harper and Row, 1990.

Johnsen, Linda. *The Living Goddess.* Saint Paul, MN: Yes International Publishers, 1999.

Jeremijenko, Valerie, Ed. *How We Live Our Yoga.* Boston: Beacon Press, 2001.

Kadestsky, Elizabeth. *First There is a Mountain.* Boston: Little, Brown and Company, 2004.

Kashyap, R. L. and S. Sadagopan, Ed. *Rig Veda Samhita.* Bangalore: Sri Aurobindo Kapali Sastry Institute of Vedic Culture, 1998.

Khanna, Madhu. *Man in Nature* Vol. 5. Edited by Baidyanath Saraswati. New Delhi: Indira Gandhi National Centre for the Arts, 1995.

Kenoyer, Jonathan Mark. *Ancient Cities of the Indus Valley Civilization.* Karachi: Oxford University Press, 1998.

Kinsley, David. *Ecology and Religion: Ecology and Spirituality in Cross-Cultural Perspective.* New Jersey: Prentice Hall.

——. *The Divine Player: A Study of Kṛṣṇa Līla.* Delhi: Motilal Banarsidaas, 1979.

Klostermaier, Klaus K. "Spirituality and Nature." in *Hindu Spirituality: Vedas Through Vedanta.* Edited by Krishna Sivaraman. New York: The Crossroad Publishing Company, 1989, pp. 319–337.

——. "Bhakti, Ahimsa, and Ecology." *Journal of Dharma 16 (July-September* 1991): 246–54.

Koelman, Gaspar M. *Pātañjala Yoga: From Related Ego to Absolute Self.* Poona, India: Papal Anthenaeum, 1970.

Larson, Gerald James. *Classical Sāṃkhya.* Motilal Banarsidass, 1969.

Larson, Gerald James, and Ram Shankar Bhattacharya, Eds. *Sāṃkhya: A Dualist Tradition in Indian Philosophy.* Princeton, N.J.: Princeton University Press, 1987.

Larson, Gerald James. "An Old Problem Revisited: The Relation between Sāṃkhya, Yoga and Buddhism." *Studien zur Indologie und Iranistik* 15 (1989): 129–46.

——— "Classical Yoga as Neo-Sāṃkhya: A Chapter in the History of Indian Philosophy," *Asiatische Studien* 53 (1999): 723–732.

Macy, Joanna, and Molly Young Brown. *Coming Back to Life: Practices to Reconnect Our Lives, Our World.* Gabriola Island, BC: New Society Publishers, 1998.

Mallinson, James. *The Kecaividyā of Ādinātha.* New York: Routledge, 2007.

Mann, R. S., Ed. *Nature-Man-Spirit Complex in Tribal India.* New Delhi: Concept Publishing Company, 1981.

Mathur, Sri Rakesh. "Can India's Timeworn Dharma Help Renew a Careworn World?" *Hinduism Today,* July 1995, 1, 9.

Miller, Barbara Stoler. *Yoga, Discipline of Freedom: The Yoga Sutra Attributed to Patanjali.* Berkeley: University of California Press, 1995.

Miller, Jeanine. *The Vedas.* London: Rider and Company, 1974.

Müller, Max. *The Six Systems of Indian Philosophy.* London: Longmans, Green and Co., 1899.

Muller-Ortega, Paul. "Tantric Meditation: Vocalic Beginnings," in *Ritual and Speculation in Early Tantrism: Studies in Honor of André Padoux*, Teun Goudriaan, Ed. Albany: SUNY Press, 1992, pp. 227–229.

Naess, Arne. 'The Deep Ecology Eight Points Revisited.' In *Deep Ecology for the Twenty-first Century*, Ed. George Sessions, 64–84. Boston: Shambhala, 1995.

———. *Ecology, Community and Lifestyle: Outline of an Ecosophy.* Cambridge: Cambridge University Press, 1989.

———. *Gandhi and Group Conflict: An Exploration of Satyagraha.* Oslo: Universitetetsforlaget, 1974.

———. "Self-Realization: An Ecological Approach to Being in the World." In *Thinking Like a Mountain: Towards a Council of All Beings,* Ed. John Seed, Joanna Macy, and Arne Naess, 19–30. Philadelphia: New Society Publishers, 1988.

Nagarajan, Vijaya Rettakudi. "The Earth as Goddess Bhu Devi: Toward a Theory of 'Embedded Ecologies' in Folk Hinduism." In Nelson, Ed., *Purifying the Earthly Body of God.* Albany: State University of New York Press, 1998.

Narayan, Vasudha. "One Tree is Equal to Ten Sons: Hindu Responces to the Problems of Ecology, Population, and Consumption." *Journal of the American Academy of Religion,* 65(2), 291–332.

Nelson, Lance E. "The Dualism of Nondualism: Advaita Vedanta and the Irrelevance of Nature." In Nelson, Ed., *Purifying the Earthly Body of God.*

———, Ed. *Purifying the Earthly Body of God: Religion and Ecology in Hindu India.* Albany: State University of New York Press, 1998.

———. "Reverence for Nature or the Irrelevance of Nature? Advaita Vedanta and Ecological Concern." *Journal of Dharma* 16 (July–September 1991): 282–301.

———. "Theism for the Masses, Non-dualism for the Monastic Elite: A Fresh look at Samkara's Trans-theistic Spirituality." In *The Struggle over the Past: Fundamentalism in the Modern World,* Ed. William M. Shea, 61–77. Lanham, Md.: University Press of America, 1993.

Niranjananda Saraswati, Swami. "Yoga and Ecology," in *Yoga (Sivananda Math),* 11(3), May 2000, pp. 14–19.

Oberhammer, Gerhard. *Strukturen Yogischer Meditation.* Vol. 13. Vienna: Verlag der Osterreichischen Academie der Wissenschaften, 1977.

Pandit, B. N. "Yoga in the Trika System," in *Specific Principles of Kashmir Śaivism*. New Delhi: Munshiram Manoharlal, 1997, 99.

Panikkar, Raimoundo. *Mantramanjari: The Vedic Experience*. Berkeley: University of California Press, 1977.

Pensa, Corrado. "On the Purification Concept in Indian Tradition, with Special Regard to Yoga." *East and West* (n.s.) Vol. 19: 194–228, 1969.

Pflueger, Lloyd W. "Dueling with Dualism: Revisioning the paradox of *puruṣa* and *prakṛti*." In Ian Whicher and David Carpenter, Eds. *Yoga: The Indian Tradition.* New York: Routledge Curzon, 2003, pp. 70–82.

Prembhava Saraswati, Swami. "The Power of the Garden," in *Yoga (Sivananda Math)*. 22–27.

Prime, Ranchor. *Hinduism and Ecology: Seeds of Truth.* London: Cassell, 1992.

———. *Vedic Ecology: Practical Wisdom for Survivng the 21st Century.* Novato, CA: Mandala Publishing (Originally published under the title Hinduism and Ecology in 1992), 2002.

Raj, A. R. Victor. *The Hindu Conection: Roots of the New Age.* St. Louis: Concordia Publishing House, 1995.

Rambachan, Anatanand. "The Value of the World as the Mystery of God in Advaita Vedanta." *Journal of Dharma* 14 (July–September 1989): 287–97.

Ravindra, Ravi. *Science and Spirit.* New York: Paragon House, 1991.

Ray, Amit. "Rabindranath Tagore's Vision of Ecological Harmony." In James, Ed., *Ethical Perspectives,* 217–40.

Rolston, Holmes, III. "Can the East Help the West to Value Nature?" in *Philosophy East and West, 37.* April 1987. 172–90.

Rukmani, T. S., Trans. *Yogavārttika of Vijñānabhikṣu: Text along with English Translation and Critical Notes along with the Text and English Translation of the Pātañjala Yogasūtras and Vyāsabhāśya.* Vol. 1: *Samādhipāda* (1981); Vol. 2: *Sādhanapāda* (1983); Vol. 3: *Vibhūtipāda* (1987); and Vol. 4: *Kaivalyapāda* (1989). New Delhi: Munshiram Manoharlal, 1981–89.

———. "Environmental Ethics as Enshrined in Sanskrit Sources." *Nidān* (Journal of the Department of Hindu Studies and Indian Philosophy, Durban) 7 (1995).

Sanderson, Alexis. "Maṇḍala and Āgamic Identity in the Trika of Kashmir," in André Padoux, Ed., *Mantras et diagrammes rituels dans l'hindouisme.* Paris: Editions du CNRS, 1986, 169–207.

Saraswati, Swami Muktibodhananda, translator. *Hatha Yoga Pradipika: Light on Hatha Yoga.* Munger, India: Bihar School of Yoga, 1985.

Seed, John. *Thinking Like A Mountain: Towards A Council of All Beings.* Philadelphia, PA: New Society Publishers, 1988.

Seed, John. "The Rainforest As Teacher," in *Inquiring Mind, 8*(2), 1992.

———.“Spirit of the Earth: A Rainforest Activist Journeys to India for Soul Renewal," in *Yoga Journal, 138* (1988): pp. 69–71, 132, 134–136.

Sen, Geeti, Ed. *Indigenous Vision: Peoples of India, Attitudes to the Environment.* New Delhi: Sage Publications, 1992.

Sherma, Rita DasGupta. "Sacred Immanence: Reflections of Ecofeminism in Hindu Tantra." In Nelson, Ed., *Purifying the Earthly Body of God.* Albany: State University of New York Press, 1998.

Shiva, Vandana. *Protect or Plunder?* London & New York: Zed Books, 2001.

———. *Earth Democracy: Justice, Sustainability and Peace.* Cambridge: South End Press, 2005.

———. *Close to Home: Women Reconnect Ecology, Health, and Development Worldwide.* Philadelphia, PA: New Society Publishers, 1994.

———. *Staying Alive: Women, Ecology, and Development*. London: Zed Books, 1989.

Singleton, Mark and Jean Byrne, Eds. *Yoga in the Modern World.* New York: Routledge, 2008.

Skolimowski, Henryk. *EcoYoga: Practice & Meditations for Walking In Beauty on the Earth.* London: Gaia Books Limited, 1994.

Sochaczewski, Paul Spencer. "The Saga of Krishna's Gardens: Can Love and Faith Heal Enviornmental Sacrilege?" *International Herald Tribune,* 18 October 1994.

Sohoni, S. Shrinivas, Trans. *Pṛthvisukta.* New Delhi: Sterling Publishers, 1991.

Srinivasan, Doris. *Religious Significance of Divine Multiple Body Parts in the Atharva Veda.* Unknown Binding, 1978.

Svatmarama. *The Hatha Yoga Pradipika.* Pancham Sinh, translator. New Delhi: Munshiram Manoharlal, 1997.

Taylor, Bron R. *The Encyclopedia of Religion and Nature.* London; New York: Thoemmes Continuum, 2005.

Thomashow, Mitchell. *Ecological Identity.* Cambridge: MIT Press, 1996.

Tigunait, Pandit Rajmani. "Our Planet, Our Selves," in *Yoga International,* 1999, September, pp. 23–29.

Tobias, Michael, Ed. *The Soul of Nature: Visions of a Living Earth.* New York: Continuum, 1994.

Tucker, Mary Evelyn, and John A. Grim, Eds., *Worldviews and Ecology: Religion, Philosophy, and the Environment.* Maryknoll, N.Y.: Orbis Books, 1994.

Vannucci, Marta. *Ecological Readings in the Veda: Matter, Energy, Life.* New Delhi: D.K. Printworld, 1994.

Varenne, Jean. *Yoga and the Hindu Tradition.* Chicago: University of Chicago Press, 1973.

Vatsayam, Kapila. "Ecology and Indian Myth." In Sen., Ed., *Indigenous vision.*

——. Ed. *Prakriti: The Integral Vision.* 5 Vols. New Delhi: Indira Gandhi National Centre for the Arts; D.K. Printworld, 1995.

Whicher, Ian. "The Integration of Spirit (*Puruṣa*) and Matter (*Prakṛti*) in the *Yoga Sūtra.*" In *Yoga: The Indian Tradition,* Edited by Ian Whicher and David Carpenter, 51–69. New York: RoutledgeCurzon, 2003.

Whicher, Ian, and David Carpenter, Eds. *Yoga: The Indian Tradition.* New York: RoutledgeCurzon, 2003.

Whicher, Ian R. "Nirodha, Yoga Praxis and the Transformation of the Mind." *Journal of Indian Philosophy*, Vol. 25: 1–67, 1997.

———. *The Integrity of the Yoga Darśana.* Albany: State University of New York Press, 1998.

Wilkoszewska, Krystyna, Ed. *Aesthetics of the Four Elements: Earth, Water, Fire, Air.* Ostrava: University of Ostrava Tilia Publishers, 2001.

Whitney, William Dwight, Trans. *Atharva-Veda-Samhit.* Delhi: Motilal Banarsidass, 2001.

Zimmerman, Francis. *The Jungle and the Aroma of Meat: An Ecological Theme in Hindu Medicine.* Berkeley: University of California Press, 1987.

CONTRIBUTORS

Christopher Key Chapple is Doshi Professor of Indic and Comparative Theology at Loyola Marymount University. He has published several books, including *Hinduism and Ecology* (with Mary Evelyn Tucker); *Jainism and Ecology; Karma and Creativity; Nonviolence to Animals, Earth, and Self in Asian Traditions; Reconciling Yogas*; and *Yoga and the Luminous: Patanjali's Spiritual Path to Freedom*. He serves on the advisory boards for the Forum on Religion and Ecology, the Ahimsa Center, and the Dharma Association of North America.

Al Collins is a clinical psychologist with a second Ph.D. in Indology. The author of a number of papers, book chapters, and one book on psychoanalysis and Indian psychological theories, his current research centers on developing a culture theory based on Indian and psychoanalytic sources and applying it to contemporary societies.

Laura Cornell, Ph.D., is an independent scholar, Kripalu Yoga teacher, and holistic health consultant. She founded the Green Yoga Association, which helped spur a national movement toward non-toxic Yoga mats, greening Yoga studios, and increased connection to the planet through Yoga practice. Laura lives in Oakland, CA.

Beverley Foulks has a B.A. in Comparative Literature from Stanford University, a M.Div. from Harvard Divinity School, and a Ph.D. in East Asian Languages and Civilizations from Harvard University. Her research interests include Asian religions, repentance, ritual studies, and comparative religious ethics.

Suzanne Ironbiter has a doctorate in history of religion from Columbia University and teaches at SUNY Purchase College and Hunter College, CUNY. Her book Devi: *Mother of My Mind* (Mapin-Lit 2006) is a personal poetic interaction with the goddess tradition in the mythology, mysticism, and philosophy of India.

Knut A. Jacobsen is professor in the History of Religions at the University of Bergen, Norway. He is the author of *Prakrti in Samkhya-Yoga: Material Principole, Religious Experience, Ethical Implications* (Peter Lang, 1999; Indian edition Motilal Banarsidass 2002). Recent books include *Kapila: Founder of Samkhya and Avatara of Vishnu* (Munshiram Manoharlal, 2008) and *Modern Indian Culture and Society* (Routledge, 2009). He is also the editor in chief of the multivolume *Brill's Encyclopedia of Hinduism* (Brill 2009–2013).

Jeffrey S. Lidke, Ph.D., is associate professor of religion at Berry College where he chairs the Interfaith Council and Asia Studies Task Force. In addition to teaching courses on world religions, ecology, yoga, and Buddhist and Hindu philosophy, Dr. Lidke also offers courses in Tai Chi and hand percussion (particularly tabla). He is the author several works on yoga and tantra, many of which are based on his extensive field research in Nepal, India, Bali, and Bhutan.

Ian Whicher earned his Ph.D. from the University of Cambridge and is currently a Professor in the Department of Religion at the University of Manitoba in Winnipeg, Canada. Dr. Whicher specializes in philosophies of India and the Yoga tradition. He is the author of several books and articles including, *The Integrity of the Yoga Darsana* (State University of New York Press), coeditor with David Carpenter of *Yoga: The Indian Tradition* (RoutledgeCurzon) and is currently engaged in a project on *Freedom and Fullness: The Reconciliation of Contemplation and Action in the Yoga Tradition.*

INDEX

Abhinavagupta, 10
Absolute, 42, 52, 54, 74, 107, 110–112, 116, 129, 132, 135
abstinence, 116
Adorno, Theodor, 88
Advaita Vedānta, 34, 59, 107, 119, 124
affliction (See kleśa), 36, 38–39, 45, 67, 71
Agarwala, V. S., 14
ahaṃkāra, 53, 59, 85–87, 89, 92–94, 101–102
ahiṃsā, 46, 52, 98, 107, 108–112, 114, 117, 120, 147, 158, 162, 180, 189
alchemy, 163
aloneness, 39, 41, 46–49, 57–58, 62, 64–65, 77, 83
animal, 16, 20, 100, 114, 139
anthropocentric, 109
aparigraha, 52, 99, 159
Āraṇyaka Yoga, 151, 155, 163, 167
åsana, 61, 100, 106, 150, 153, 159
ascetic, 61, 74, 105, 112–113, 118
Aṣṭāṅga-yoga, 42, 56
asteya, 52, 99, 158
Atharva Veda, v, 3, 9, 13–16, 21–23, 30–31, 186–187
Ātman, 5, 30, 107, 124, 152
Aum, 71, 89–90
Aurobindo, Sri, 14–15, 21–22, 172, 179
austerity, 99
avidyā, 36, 38–40, 55, 58, 63, 66
avyakta, 86
awareness, 3, 20, 28, 30, 39, 46, 49, 69, 76, 112, 131, 133, 138–140, 145–146, 148–149, 151, 153, 157, 159, 163–164, 166–169
beauty, 30, 101, 104, 113, 127, 139–140, 157, 164, 185
Bhabha, Homi, 92–95, 172
Bhagavad Gītā, 1, 2, 5, 6, 106, 108, 118–119, 123 124, 143, 152, 162, 178
bhakti, 91, 106, 118–119, 151, 153–154, 167, 168, 180
Bhakti Yoga, 167, 180
bandha, 128, 133
Bhikṣu, Vijñāna, 34, 45, 55, 59
bhoga, 84, 85, 116, 125
bīja, 84, 85, 116, 125
biosphere, 147
bindu, 131
Bodhisattva, 80, 90, 91
bondage, 38, 66, 77, 88, 102, 115, 126, 138
Brahmā, 5, 16, 19, 28, 106, 107, 135
brahmacarya, 52, 99
Brahman, 15, 17, 19, 27, 67, 78, 106–107, 152
brahmanda, 133

breath, 5, 29, 61, 71, 100, 101, 148, 151, 153, 154, 156–159
Buddha, 2, 90
buddhi, 53, 80, 83–84, 86, 87, 88–89, 93, 101, 102, 116
buddhīndriya, 25, 101
Buddhist, 7, 52, 70-71, 80, 90, 191
Buddhism, 7, 67, 68, 90, 180
cakra, 89, 131, 136, 142
Callicott, J. Baird, 2, 11
Carpenter, David, 2, 11, 173
cessation, 35–36, 40–41, 44, 47, 49–50, 53, 55, 59, 61–66, 68–75, 77–78
chastity, 52
citiśakti, 39, 54, 88–89
citta, 35–36, 39–40, 53,55, 60, 71, 84
citta-vṛtti, 47, 56, 61, 63–64, 65, 68–69, 71, 73, 75–78, 101
clarity, 46, 54, 66, 98, 102, 166
Cohen, Leonard, 91
community, 14, 20, 118, 120, 139, 145–146, 148, 150–151, 160, 163–165, 167
compassion, 46–47, 54, 71, 158, 161–164
consciousness, 4, 10, 23, 30, 33, 36, 38–40, 43–46, 48–50, 53–54, 62–64, 65, 67, 69–71, 76–77, 83, 85, 87–89, 91–92, 97–98, 112, 117, 124–125, 128–134, 140, 148, 151–152, 157, 160, 160, 162, 165, 167, 173
contentment, 99, 103
cosmos, 3, 5, 9, 19, 132, 135, 139, 147
Dayal, Har, 67, 80
darśana, 33, 38, 43, 44, 51, 53, 62, 67–68, 79–81, 106, 124, 167, 192
DasGupta, Rita, ii, vii, 1, 2, 126–127, 137, 141, 185
Deep Ecology (Movement), 6, 105, 107–109, 111–112, 114, 116–119, 123, 141, 141, 174, 178, 181
Deepak, Adarsh, vii
depression, 103
detachment, 41, 46, 108
Deva(ta), 13, 18–20, 141
Devī, 14, 27
dhāraṇa, 61, 95, 101, 159
Dharma Association of North America, ii, vii, 189
Dharma Megha Samādhi, 5, 62, 97–98
dhyāna, 61, 95, 159
dīkṣā, 131
discernment, 41, 56, 70, 84, 97
disharmony, 43, 105, 108–112, 117, 120
dispassion, 44, 56, 68–71
dissolution, 37, 38, 42, 130–131
divinity, 116, 128, 131, 138, 152, 159, 163, 190
draṣṭṛ, 38–39, 46
dualism, 34, 47, 62, 80, 107, 109, 128
duḥkha, 37, 38, 41, 54, 60
Dwivedi, O. P., 8, 11, 31, 176

earth, v, 5, 8, 10, 13, 16, 19, 23–31, 98, 101, 104, 116, 121, 126–127, 129, 135, 138, 140–141, 145, 147, 150, 153–154, 159, 161–164, 166, 173–177, 179–180, 182, 185–187, 189
ecofeminism, 126, 185
ecology, v, 6–7, 9–11, 21, 104–105, 112, 116, 118, 137–141, 145, 147, 165–166, 173, 175–183, 185–187, 189, 191
ecological, 1–3, 6, 8, 10, 11–14, 21, 23–24, 95, 97, 103, 108, 110–113, 116–117, 126–128, 136–137, 139, 145–146, 148–152, 156, 158, 160, 163, 165, 171–174, 181–182, 184, 186–187
ecosophy, 127, 141, 178, 181
Ecosophy T, 10, 106–107, 141, 178
ego, 45, 57, 59, 94, 102, 107, 180
eightfold-path, 111
Eliade, Mircea, 20–22, 52, 56, 87, 176
emptiness, 88, 92, 94–95
environmentalism, 29, 104, 109, 117, 119, 123, 171
equanimity, 60, 71, 160
ethical, 2, 7–8, 35, 46–47, 50, 98, 102, 108, 120, 123, 147, 158, 161, 171, 175–176, 178–179, 184, 191
ethics, v, 2, 5–6, 9–10, 30, 47, 49, 52, 97–98, 103–104, 110, 112, 117, 184
faith, 3, 6, 70, 186
feminine, 2, 6, 9–11, 21, 125–126, 165
fire, 8, 10, 16, 19, 23, 26–27, 91–92, 129, 131, 135, 140, 150, 165-166, 177, 187
Flood, Gavin, 67, 80, 142, 177
fragrance, 23, 25–26, 101
Frawley, David, 8, 21, 177
freedom, 4, 11, 35, 37, 39, 50, 56, 62, 64–65, 74, 77, 79, 81, 83–84, 88–89, 95, 103, 110, 114, 115, 123, 176, 181, 189, 192
Freud, Sigmund, 87
Ganges, 136
Gāyatrī Mantra, 155, 166
Gheraṇḍa Saṃhitā, 100–101, 177
God, 3, 8, 11, 23, 26–27, 59, 99, 128, 129–130, 132, 134, 142, 153–154, 173–176, 183–188, 185
Goddess, 9, 91, 126, 129, 132–135, 132, 138, 154, 179, 190
Gonda, Jan, 14, 21, 110, 120
Green Yoga (Association), vii, 6, 9, 104, 145–163, 165–168, 174, 177
Grim, John, vii, 173, 186
guṇa, 30, 36–41, 44, 54–55, 83, 86, 88, 110

Gupta, Sanjukta, 129, 141–142, 178
Haberman, David, 8, 118
Halifax, Joan, 16, 21, 178
harmony, 40, 49, 110–111, 112, 116–117, 140, 158
Haṭha yoga, 3, 100, 128, 145, 154, 156–158, 162, 165, 184
Haridbhadra, 3
heaven, 5, 13, 19, 110, 116
Hemacandra, 3
Himalaya, 3, 143
Himavat Khaṇḍa, 136, 143
Hiraṇyagarbha, 17, 18
hologram, 135
hymn, 3, 9, 13–15, 17, 20, 22, 23, 27, 30, 31, 136, 172, 177
ignorance, 36, 38–41, 43, 49–50, 55, 58, 59, 66, 73, 85, 87, 102
identity, 33, 36–42, 44–45, 47–50, 53, 57–60, 64, 68, 72, 77, 85, 87, 93, 102, 109, 110, 112, 123, 142, 184, 186
individuality, 44, 164
Indra, 3, 15, 18, 19, 23, 26–27
Indus Valley, 2, 16, 179
initiation, 67, 128, 130–131, 139
insight, 6, 11, 17, 30, 35, 43, 46, 48, 83, 84, 86–91, 97, 102, 112, 124–125, 127, 162, 164
iśvara, 61, 71, 99, 141
Iśvarakrishna, 4, 34, 58
Jacobsen, Knut, 6, 105, 118–121, 123–124, 137, 141, 178, 191
Jaina, 3, 52
James, George A., 8, 179
Jayakar, Papul, 14–16, 21
Jina, 2
jīva, 107, 138
jīvanmukti, 33, 45, 58–59
jñāna, 35, 48, 83, 85, 89, 92, 151
Jñana Yoga, 151–152, 167
Kabir, 3
kaivalya, 4, 39–51, 54–56, 62, 64–65, 70, 77, 83–89, 93–94, 97–98, 111, 117, 184
kāmadhenu, 24
Kapoor, Anish, 90, 92–95, 172
karma, 4, 45, 58–59, 66–68, 74–75, 78, 97, 174
Karma Yoga, 59, 105–107, 117–118, 150–151, 160, 167
Kālidāsa, 113–114, 117, 121
Kapani, Lakshmi, 66
Kapila, 4, 191
karuṇā, 46
Kaula, 134, 143
Khanna, Madhu, 14, 21
Khyentse, Dilgo, 138
knowledge, 17–18, 20, 34–35, 40, 42, 44–45, 48–50, 55–56, 71, 74, 104, 111–112, 114–116, 126, 128, 134, 151, 167
Klostermaier, 57, 90, 95
kripalu, 9, 146, 168, 712, 175
Krishna, 5, 8, 54, 118–119, 161, 186
kumārī, 132
kuṇḍalinī, 130–131, 135

liberation, 4, 10, 30, 35, 37, 41–42, 45, 49, 54–58, 66, 85–86, 97, 112, 115, 123–124, 126, 137, 165
liṅgam, 16
loka, 130
Lord of Yoga, 61, 89
love, 91–92, 108, 114, 150, 154–155, 159, 161–163
luminosity, 45
Mahābhārata, 2, 120
maṇḍala, 14, 127–128, 134, 139–142
Marxism, 88
masculine, 165
materiality, 85, 87, 91, 110–112, 116, 126, 165–166
materialism, 7
mātṛkā, 130, 133, 135
matrix, 6, 29–30
māyā, 2, 138
mantra, 142, 155–156
memory, 53, 61–63, 65–66, 72, 75, 102
Miller, Barbara Stoler, 62, 70–72, 79, 81, 84–85, 95
Miller, Jeanine, 14, 20–21
mind, 5, 14, 19, 30, 33, 35–37, 39–42, 45, 48–50, 53–55, 58–60, 63, 65–66, 69–71, 77, 84, 101, 104, 115, 124, 157–159, 167, 185, 187, 190
Miśra, Vācaspati, 34, 110
Mohenjodaro, 2
monism, 109
mother, 24–30, 31, 133
Mount Meru, 128–129
Müller, Max, 41
Nanak, Guru, 3
nāḍī, 17
Naess, Arne, 6, 10 , 105–109, 117–120, 181
Nāth Siddha, 128
nature, 1–2, 4, 5, 7–10, 13–14, 16–17, 19–21, 23, 25, 30, 33, 39–43, 45, 47, 52–55, 57, 59, 65, 74, 76, 78, 83–86, 88–91, 97, 99–103, 107–117, 120, 124–126, 129–130, 133–134, 139–141, 147, 150–153, 155–156, 158–159, 162–165, 168, 171–177, 179–186
nectar, 134
Nelson, Lance, 1, 8, 11, 104, 137, 182
nirbīja, 55, 69, 97
nirbīja-samādhi, 41, 62–64, 66, 71–72, 76
nirodha, 35–36, 46–48, 53, 55, 57, 62–63, 187
nirvāṇa, 87, 90, 106
Nityāṣoḍaśikārṇava, 131, 138, 142
niyama, 61, 102, 150, 158–159
nonviolence, 46, 52, 98, 101, 103–104, 158, 162, 173, 174
obstacles, 2, 41, 71, 73
Panikkar, Raimundo, 14, 21–22, 183
Paramārtha, 128
personality, 41, 44, 92, 102

Pflueger, Lloyd, 64, 80, 183
prakṛti, 2, 4–5, 9, 30, 33–59, 62–65, 69, 72, 74–80, 83–94, 97, 110–112, 120, 124, 133, 178, 183, 187, 191
Paramaśiva, 130, 134
Patañjali, 2–3, 6, 11, 33–35, 37–39, 42–43, 46, 48–49, 51–54, 58, 60, 61–64, 66–68, 70 76, 78–81, 95, 97–104, 120, 150, 158, 160, 172, 176–177, 181, 189
Paśupati, 2
peace, 26, 106, 110, 113–114, 150–151, 155, 159, 161–162, 165, 185
plant, 2, 26, 75, 102, 115, 139, 146–150, 154, 156–157
poet, 13–21, 58, 99, 113, 128
pollution, 1, 24, 28, 103, 118, 171, 178
posture, 61, 100, 106, 113, 149–150, 153, 156, 159, 168
potter's wheel, 56, 89
Poussin, Louis de la Vallée, 68, 80
Prajapati, 15, 17, 19
prajñā, 43, 61, 63, 70, 72
prāṇa, 61, 101, 150, 168
prāṇāyāma, 61, 100, 156–157, 159
pratiprasava, 35, 37–40, 54, 84, 86, 88–89
pratyāhāra, 61, 101, 159
Prime, Ranchor, 8
Pṛthivī Sūkta, 3, 23–31
Pumān, 19
Purity, 4, 36–37, 40, 46, 48, 99
puruṣa, 4–5, 15, 17–19, 30, 33–59, 62–69, 72–80, 83–96, 97, 110–112, 116 , 124, 183, 187
puruṣārtha, 54, 84–85, 87–88, 90, 94–95
Radhakrishna, Sarvapalli, 108, 119
Rāja, Bhoja, 34, 41
Rāja Yoga, 158, 162, 167
rajas, 30, 36, 44, 55, 87, 110
Rao, K. L. Seshagiri, 14, 21
realism, 4, 142
rebirth, 68, 103, 114
reflection, 24, 48, 123–126, 129, 162, 178, 185
Renou, Louis, 15
ṛta, 16, 18, 43
ṛtaṃbharā, 43
Rukmani, T. S., 14, 21, 55, 184
renunciation, 6, 10, 33, 105
restraint, 3, 7, 52, 99, 102–103
sabīja, 63, 65, 69, 72
sacrifice, 18, 20, 27, 46, 98, 108, 142
sādhaka, 128, 130–131, 133–135
sādhana, 37, 71, 15, 128, 130–131, 133, 142
Śakta, 126–127, 129
Śakti, 9, 39, 89, 127, 129–131, 140
salvation, 35, 120
samādhi, 5, 36, 40–41, 44–45, 47–48, 52–53, 55–57, 59, 61–66, 70–72, 76, 83–95, 97–98, 101, 124, 159, 172, 184
sāman, 19

Sāṃkhya, 2, 4, 9, 25, 34, 37, 39, 52–53, 55–58, 62, 79, 83–85, 92–95, 105, 108–112, 114–117, 120–121, 172, 178–180, 191
Sāṃkhya Kārikā, 4, 25, 34, 55, 83–85, 92
saṃsāra, 41–42, 45, 56, 66, 70, 123
saṃskāra, 4, 9, 40, 55–56, 61–81, 97–98, 102
saṃyoga, 39, 42, 44, 46, 48, 54, 63
Śāṅkara, 52, 59–60
santoṣa, 99
satkaryavāda, 4
sattva, 30, 36, 40–41, 44, 54–55, 87, 110
satya, 16, 52, 98, 158
Śauca, 99
seed, 36, 55, 63, 65–66, 69, 71–73, 75, 133, 191
Self-Realization, 57, 106–108, 111, 114, 118, 123–124, 177, 181
sexuality, 132
Sherma, Rita Dasgupta, 1–2, 126–127, 137, 185
Shiva, Vandana, 8, 11, 125–126, 137, 141, 185
siddha, 83, 128, 130–131, 171
Sikh, 3
Śiva, 16
Śiva Samhitā, 128, 141
Skambha, 13–20, 23, 30
skandha, 67
Skolimowski, Henryk, 8, 10, 112, 185
Smith, J. Z., 127
solitude, 4, 109
soma, 27
soul, 4–5, 110, 112, 129, 138, 150, 155–157, 161–163, 173, 185–186
species, 1, 9, 74, 98, 102, 107, 114, 138, 158, 160
Srinivasan, Doris, 15, 21
srāddha, 16, 70
Śri Vidya Tāntrika, 134
sruti, 14
suffer(ing), 7, 54, 73, 83, 85–88, 92, 102, 108–109, 111–112, 123–124, 138, 166
Śūnya, 84–85, 88, 94
sustainable development, 116
Taber, J., 49, 60
tamas, 30, 36, 44, 55, 87, 110
tanmātra, 25, 101
Tantra, 1–2, 6, 10, 91, 123, 126–142, 151, 165, 167, 177, 185, 191
Tantric, 9, 91, 127–133, 136–141, 181
Tāntrika, 127, 132, 134–135
Tapas, 16, 18–19, 61, 99
Tattva, 34, 88, 130
Taylor, Bron, 8, 186
technology, 99, 115, 132, 135
transcendence, 2, 7, 40, 46, 49, 97, 132, 134, 137
transcendent, 37, 39, 46, 48, 64, 112, 114, 136, 165–166

tree, 13–16, 19–20, 113, 124, 127, 140, 149, 153, 155–158, 164, 182
Trika, 134, 142–143, 183–184
truth, 7, 13, 16, 19, 43, 52, 98, 105, 158, 162, 183
truth-bearing, 43
Tucker, Mary Evelyn, 8, 11, 21, 121, 194
turīya, 87
Upaniṣad, 2, 5–6, 9, 26, 30, 61, 87, 89, 100, 152
utilitarian, 7, 115–116
Vaishnava, 8
Vālmīki, 114
Veda, 3, 9, 13–17, 19–22, 23, 26–27, 30–31, 57, 61, 120, 154, 172, 177, 179–181, 186–187
Vedānta, 2, 34, 57, 59, 105, 107–108, 119, 124, 175, 180, 182–183
Vedic, 1, 3, 9, 14–19, 21–22, 23, 29, 66, 155, 179, 183
vegetarian, 111, 120
vīrya, 70
visarga, 130
Vishnu, 23, 26–27, 191
viveka, 41, 56, 70, 84, 90, 97, 111
viyoga, 42, 45
vrata, 16
Vrindivan, 86
Vṛtra, 3, 26
vṛtti, 35, 36, 40, 44–45, 53, 56, 61, 63–66, 68–69, 71–73, 75–78, 101
Vyāsa, 34, 39, 42–45, 50, 53, 61, 63, 66, 69–70, 103, 172
White, Lynn Jr., 7–8, 23
Womb, 18, 101, 131, 133
Women, 1–2, 11, 19, 26, 141, 185
world-denying, 33
world-negating, 8
World Yoga Center, 149
worship, 8, 15, 17–18, 29, 71, 118, 128, 130–131, 153
Yajña, 17, 19
Yama, 52, 61, 98, 102, 150, 158
Yantra, 131–134
Yoga Sūtra, 3, 6, 11, 33–59, 61–81, 84–85, 90, 95, 97–98, 100, 104, 172, 177, 181, 187
Yogāvasiṣṭha, 3, 10
Yogadṛṣṭisamuccya, 3
Yogaśāstra, 3
Yogeśvara, 89
Yonī, 131, 133